一天始於下班後

妥善運用你的待機時間

除了上班開機、睡覺關機，人一天中最寶貴的「待機時間」該拿來做些什麼呢？

有些人會安排很多計畫，把每個空暇的縫隙都塞得滿滿的，看起來多有意義、多充實啊！

但「理想很豐滿，現實很骨感」，

我們經常會在要踏出去做的那一刻因為各種原因打消念頭。

所以，我們要學習的不只是懂得利用時間投資自己，

更要跨出名為「開始行動」的第一步——

阿諾・班奈特——著

（Arnold Bennett）

序
早起一小時，贏得一輩子

　　本書發行至今，我已收到許多不論是書評或是讀者感想的信件，所幸的是大多數的人對本書都給予肯定而積極的認同。其中也有不少書評已經在媒體上發表，甚至有些文章其篇幅之長還超越本書的字數。

　　雖然有些讀者建議，他們希望本書的語氣能再更為嚴肅一點，才能更加強化時間管理的重要性，但我並不會因此而有所改變，因為文字的調性輕鬆並不會降低文章所要彰顯的意義，只要能讓大部分的讀者感到有益，不論意見是否切中要害的評論或無關緊要的瑣事，我都會欣然接受。

　　唯獨有一封信函卻讓我耿耿於懷。雖然不是公開發表的專家評論，但這封來自於平凡讀者的誠懇文字，卻引起我更多的關注及重視，因為信中有一段真實而貼切的描述給了我們更多的啟示：

「每一天，他總是盡可能地採取拖延戰術才肯開始進行工作；就連平常的例行作業，他也很難盡力完成。總之，當工作結束的時刻才是他苦悶的日子裡最大的救贖。看來他在大部分的時候，不但對自己的事業失去了最初的熱情與動力，即使是最好的情況也只能盡量做到不去厭惡它。」

我一直都堅信，大部分的人對自己的工作總是懷抱著高度的熱愛及關注。這些人不論階級及身分，包括有位高權重的企業家，或是前途看好的青年才俊，甚至是在社會底層的勞工階層。他們不但恪守本分及時間的分際，也不會逃避應盡的責任。總之，對於工作，大家總是盡其所能地付出一切，直到心力交瘁。

過去，我曾有機會到倫敦及其他一些大城市經商，但其中有好幾年的時間，事業卻都處於低潮，幾乎要令自己失去自信。所幸當時因緣際會，認識一群和我有相同境遇的朋友，曾溫暖地給予我鼓勵。更讓人感動的，是他們對自己的工作不僅懷抱著堅持不變的信念，並始終努力不懈地投注最大的熱情，全力以赴。

　　但遺憾的是，就現今的社會狀況看來，像他們這樣在逆境中生存，還可以感到幸福，並且樂在工作上的人們（說不定，他們還自認為比我們看到的還要快樂），畢竟只是屬於「少數族群」。

　　我們當然也相信，大多數人對於自己的工作的確擁有極高的理想及期待。只是在這一群人中，仍有大部分的人是不會為了工作就將自己的能量耗費殆盡，更別說還有人是不但早就對工作失去興趣，甚至可能到了厭倦的地步。因此別說他們根本沒有絲毫的企圖心，就連對工作應盡的職責，恐怕他們也只願意做到最低限度吧！

　　因此，相對於那些始終樂在工作的人來說，對工作的付出他們顯然應該獲得更多的尊重。若要比擬這群朋友們所面對的高難度，或許可用我的一位筆友所親身經歷的描述更為貼切：

「我何嘗不想和那些樂在工作的朋友一樣呢？我當然也想對自己的工作做出更好的成績，甚至超越夢想的成就！但實際的情況是，當我將工作完成回家後，我竟發現，疲憊不堪的身心已經讓生活中的任何事都引不起我的一點興致了。」

或許，我們可以由此印證一件事：當一個人能將熱情投注於自己有興趣的事時，至少他們不會感到終日渾渾噩噩，總是對生命感到失望及悔不當初。所以，他們是不需要有人嘮嘮叨叨、耳提面命地來指導他們怎麼「過日子」！

因為，在一天二十四小時裡，亦即在他們一天的生活中，不但可說是很少出現虛度光陰的時候，甚至在工作期間，他們也總是展現出最高意志，始終活力充沛地進行著任務。

相對於那些浪費的時間來說，即使有人每天只付出築夢踏實的八個小時，也比不知所謂而過了十六個小時的人，他們的生命更顯得有價值吧！本書正是為了這一群朋友而誕生──不論是在職場上或是生活中都感到索然無味的人。

當然，那些樂在生活的朋友或許會說：「即使，我對工作的要求比別人來得多又高，但我還是想要做得更完美；即使，我已經很認真地過日子了，但我還希望能生活得更有意義。但問題是，我發現自己好像已經無法在工作之餘抽出更多的時間來完成更多的夢想了？」

是的！為了因應可能有人有這樣的需求，本書所提供的建議，或許也能讓用心過生活的讀者，意外地獲得更多的時間。

事實上在著手寫書之前，我早就預料到本書對於那些急於有更多時間可利用的大忙人來說，其吸引力可能比本書真正的目標讀者（不知為何而忙或終日無所事事的人們），要來得更為迫切需要吧！

一旦一個人真正體驗過生活的樂趣，他將會發現自己需要更多的時間。正如貪睡的人，總以為天下最難辦到的事就是早起。或許有人會質疑地說：「每天為了生活，都已經快要精疲力盡了，怎麼可能有多餘的時間來執行這些事情呢？」

　　但是我還是相信，事在人為。因為就算是每天嚷嚷時間不夠用的工作狂，甚至是為了生活而不停工作的勞動者，只要願意空出些許時間，來進行其中的一些建議，將會發現生活品質有了極大的改變。

　　至於，要如何才能找出時間來培養習慣呢？一般來說，若是以工作一整天後的晚間來練習，的確會力不從心，比較容易產生無以為繼的情況。因此，不妨多利用上班前上午的這個時段。一週五天，每個人的時間其實都一樣多，再不然，你一定可以從一個星期裡抽出至少三個晚上的空檔。

　　好吧！或許有人會抱怨地說：下班後都已經夠疲憊了，此時你通常只想好好放鬆，實在無法，也不想再進行任何工作以外的計畫。那我也會坦白地說：如果做了八個小時的工作，你都這麼累了，那剩餘的十六個小時，你不就都要用來補充體力嗎？其實，工作只是你生命中的一部分而已，有必要如此耗損掉你大部分的生命能量嗎？

　　因此假使白天的工作已經讓你感到如此疲憊不堪的話，那你可要仔細思量現在的生活，是不是已經出現危機了？重新思考人生的重心，並調整出均衡的生活步調，將是你刻不容緩的選擇。那麼，究竟要如何才能充分地利用我們的時間，並樂活人生呢？

　　首先，盡可能地在一天之中，提早啟動我們對生命展現熱情的時刻。亦即，**改掉不到最後一刻不起床的壞習慣，並且盡量早起。**一開始或許會因此感到睡眠不足，但只要能夠持續一段時間後，你將發現自己不僅不會再哈欠連連、精神不濟，身心也會比以往更加健康。另外，有人或許會因為早起干擾到家人的作息而感到困擾？別擔心，只要一旦下決心進行，任何問題都可以找到解決之道。

　　至於有人問，睡眠的時間要多久才算充足呢？是否一定要八小時呢？或九小時？其實，這只是習慣養成的問題。更重要的是別成了「懶散」的最佳藉口。就像那些天天在一大清早就準備要耕耘的農夫們，或是曙光初露就早已奔波在路上的客運司機們，他們的睡覺時間可是比我們還要來得更少，但仍然能精神奕奕地進行工作。

此外我也曾經請教過醫生，關於睡眠時間的見解。在倫敦，一名有二十四年行醫經驗的醫生，給了我一個直接而幽默的說法：「喔！那些賴床的人都是傻瓜！」

他認為，假使有十個人能習慣早起，那其中至少會有九個人，不僅會因此享有更健康的身體，當然，他們的生活品質也會比一般人的更為提高。這一看法顯然與其他許多的醫生是相同的。但唯一例外的情況，則是睡眠時間對於那些正在成長期中的孩子們來說，可能就需要得比成人還要更多。

因此建議讀者，一開始不要過於匆忙，可先嘗試提早一個小時起床，接著再試驗一個半小時，然後逐漸提早。因此今後在晚間，就必須要盡早上床睡覺。只要經過一段時間後，你將會發現，**提早一個小時起床後所能完成的事，遠比熬夜做了兩個小時以上的效率更好**。

但遺憾的是，不管我如何苦口婆心地說著早起的好處，總還是有人可以提出許多無法早起的理由，如早餐不吃無法工作啦！一定要喝現煮咖啡，才能保持清醒啦！若沒有人準備⋯⋯等等。所謂「工欲善其事必先利其器」，當然啦，完成事情的必要條件若能越齊全，百分之百達成的機率自然越高。更何況對某些人來說，若連基本需求都無法滿足的話，不但沒有堅持下去的動力，恐怕就連開始都是非常困難的。

但別忘了！**一個人面對人生的態度，正決定了他將來的成就高度**。誠如有人說，要端詳一個人是否不尋常，就看他能否在不尋常的時刻中品茗。因此，對一個下定決心要實踐的人來說，只要能邁向成功之道，縱然有天下人都無法克服的問題，他依然會排除萬難，勇往直前。

阿諾・班奈特

Preface

This preface, though placed at the beginning, as a preface must be, should be read at the end of the book.

I have received a large amount of correspondence concerning this small work, and many reviews of it—some of them nearly as long as the book itself—have been printed. But scarcely any of the comment has been adverse. Some people have objected to a frivolity of tone; but as the tone is not, in my opinion, at all frivolous, this objection did not impress me; and had no weightier reproach been put forward I might almost have been persuaded that the volume was flawless! A more serious stricture has, however, been offered—not in the press, but by sundry obviously sincere correspondents—and I must deal with it. I anticipated and feared this disapprobation. The sentence against which protests have been made is as follows:

"In the majority of instances he [the typical man] does not precisely feel a passion for his business; at best he does not dislike it. He begins his business functions with some reluctance, as late as he can, and he ends them with joy, as early as he can. And his engines, while he is engaged in his business, are seldom at their full effort."

I am assured, in accents of unmistakable sincerity, that there are many business men--not merely those in high positions or with fine prospects, but modest subordinates with no hope of ever being much better off--who do enjoy their business functions, who do not shirk them, who do not arrive at the office as late as possible and depart as early as possible, who, in a word, put the whole of their force into their day's work and are genuinely fatigued at the end thereof.

I am ready to believe it. I do believe it. I know it. I always knew it. Both in London and in the provinces it has been my lot to spend long years in subordinate situations of business; and the fact did not escape me that a certain proportion of my peers showed what amounted to an honest passion for their duties, and that while engaged in those duties they were really *living* to the fullest extent of which they were capable. But I remain convinced that these fortunate and happy individuals (happier perhaps than they guessed) did not and do not constitute a majority, or anything like a majority. I remain convinced that the majority of decent average conscientious men of business (men with aspirations and ideals) do not as a rule go home of a night genuinely tired. I remain convinced that they put not as much but as little of themselves as they conscientiously can into the earning of a livelihood, and that their vocation bores rather than interests them.

Nevertheless, I admit that the minority is of sufficient

importance to merit attention, and that I ought not to have ignored it so completely as I did do. The whole difficulty of the hard-working minority was put in a single colloquial sentence by one of my correspondents. He wrote: "I am just as keen as anyone on doing something to 'exceed my programme,' but allow me to tell you that when I get home at six thirty p.m. I am not anything like so fresh as you seem to imagine."

Now I must point out that the case of the minority, who throw themselves with passion and gusto into their daily business task, is infinitely less deplorable than the case of the majority, who go half-heartedly and feebly through their official day. The former are less in need of advice "how to live." At any rate during their official day of, say, eight hours they are really alive; their engines are giving the full indicated effort The other eight working hours of their day may be badly organised, or even frittered away; but it is less disastrous to waste eight hours a day than sixteen hours a day; it is better to have lived a bit than never to have lived at all.

The real tragedy is the tragedy of the man who is braced to effort neither in the office nor out of it, and to this man this book is primarily addressed. "But," says the other and more fortunate man, "although my ordinary programme is bigger than his, I want to exceed my programme too! I am living a bit; I want to live more. But I really can't do another day's work on the top of my official day."

The fact is, I, the author, ought to have foreseen that I should appeal most strongly to those who already had an interest in existence. It is always the man who has tasted life who demands more of it. And it is always the man who never gets out of bed who is the most difficult to rouse.

Well, you of the minority, let us assume that the intensity of your daily money-getting will not allow you to carry out quite all the suggestions in the following pages. Some of the suggestions may yet stand. I admit that you may not be able to use the time spent on the journey home at night; but the suggestion for the journey to the office in the morning is as practicable for you as for anybody. And that weekly interval of forty hours, from Saturday to Monday, is yours just as much as the other man's, though a slight accumulation of fatigue may prevent you from employing the whole of your "h.p." upon it. There remains, then, the important portion of the three or more evenings a week. You tell me flatly that you are too tired to do anything outside your programme at night. In reply to which I tell you flatly that if your ordinary day's work is thus exhausting, then the balance of your life is wrong and must be adjusted. A man's powers ought not to be monopolised by his ordinary day's work. What, then, is to be done?

The obvious thing to do is to circumvent your ardour for your ordinary day's work by a ruse. Employ your engines in something beyond the programme before, and not after, you employ them on the programme itself. Briefly, get up earlier in the morning. You say you cannot. You say it is impossible for you to go earlier to bed of a night--to do so would upset the entire household. I do not think it is quite impossible to go to bed earlier at night. I think that if you persist in rising earlier, and the consequence is insufficiency of sleep, you will soon find a way of going to bed earlier. But my impression is that the consequences of rising earlier will not be an insufficiency of sleep. My impression, growing stronger every year, is that sleep is partly a matter of habit--and of slackness. I am convinced that most people sleep as long as they do because they are at a loss for any other diversion. How much sleep do you think is daily obtained by the powerful healthy man who daily rattles up your street in charge of Carter Patterson's van? I have consulted a doctor on this point. He is a doctor who for twenty-four years has had a large general practice in a large flourishing suburb of London, inhabited by exactly such people as you and me. He is a curt man, and his answer was curt:

"Most people sleep themselves stupid."

He went on to give his opinion that nine men out of ten would have better health and more fun out of life if they spent less time in bed. Other doctors have confirmed this judgment, which, of course, does not apply to growing youths.

Rise an hour, an hour and a half, or even two hours earlier; and--if you must--retire earlier when you can. In the matter of exceeding programmes, you will accomplish as much in one morning hour as in two evening hours. "But," you say, "I couldn't begin without some food, and servants." Surely, my dear sir, in an age when an excellent spirit-lamp (including a saucepan) can be bought for less than a shilling, you are not going to allow your highest welfare to depend upon the precarious immediate co-operation of a fellow creature! Instruct the fellow creature, whoever she may be, at night. Tell her to put a tray in a suitable position over night. On that tray two biscuits, a cup and saucer, a box of matches and a spirit-lamp; on the lamp, the saucepan; on the saucepan, the lid—but turned the wrong way up; on the reversed lid, the small teapot, containing a minute quantity of tea leaves. You will then have to strike a match—that is all. In three minutes the water boils, and you pour it into the teapot (which is already warm). In three more minutes the tea is infused. You can begin your day while drinking it. These details may seem trivial to the foolish, but to the thoughtful they will not seem trivial. The proper, wise balancing of one's whole life may depend upon the feasibility of a cup of tea at an unusual hour.

目次

第一章
每一天都是生命的奇蹟

雖然大家都說：「時間就是金錢。」但是這句話，
其實還不足以說出時間的力量究竟有多重要。
因為它个只是可以金錢衡量，
事實上，時間可以創造出比金錢更重要的財富。

「唉，我不是告訴過她要好好理財嗎！這下子又在到處借錢了，真是標準的月光族。太遜了，不管是社會地位或是工作收入，都不會比人差，再說薪水就算扣除基本開銷也夠花用一整個月了，何況她還算是不太浪費的人，卻經常這樣鬧錢荒，真是匪夷所思，我看八成是理財白痴吧！」

「想不到他經濟狀況也不怎樣嘛！雖然那間房子外觀看起來是不錯，裡面的裝潢卻叫人不敢恭維地糟透了。穿著打扮嘛，也算有點牌子，只是陳舊不堪的配件，卻是與時尚格格不入。再說，既然要用名貴的瓷盤來盛料理裝面子，菜色卻如此『家常』；更讓人看不下去的是用有裂痕的咖啡杯給客人飲用高級紅茶，看來他是入不敷出了。一定是因為這樣虛榮亂買東西所造成的吧！真是不會理財，若是他的收入給我，我的生活肯定可以比他現在更好！」

看到有人因為自作自受成了敗家子（女）時，你是不是曾經和大多數人一樣暗自竊笑，還忍不住地犯了「好為人師」的癮頭，而藉機說教一番呢？

　　大家都有自己的一套理財見解，尤其是看到不知節制地購物的愛面族時，許多人在不知不覺中都會成為財經節目的名嘴。就像是人人都成了意氣風發的企業總裁，剎那間，只聽到長篇大論滔滔不絕於耳，這真是令人驕傲的時刻啊！縱然只有短短的幾分鐘……

　　由於商業的快速發展，間接地帶動了社會風氣的轉變，近幾年來，不論是平面媒體或各式各樣的電視節目，總是充斥著許多生活理財的話題，諸如教人如何投資、如何節省日常開支……等。此外，從讀者或觀眾所引發的熱烈迴響中，更可見財富對一般人的吸引力之大。甚至在最近的新聞報導中，經常可見為了討論一個人一年度的開銷，究竟可以花費或節省到什麼樣的程度，而引來各方專家達人唇槍舌劍的激烈辯論。

　　在高唱拚經濟的年代，如「怎樣用三千元過一個月」之類的文章如雨後春筍，但教人「如何妥善運用二十四小時」之類的著作，卻似乎成了不合乎潮流需要的過去式話題，而逐漸被人們所遺忘。

雖然大家都會說：「時間就是金錢。」但是我卻認為這句話，其實還不足以說出時間的力量究竟有多重要。因為它不只是可以金錢衡量，事實上時間可以創造出比金錢更重要的財富。正如我們用時間和老闆換得了這個月的薪水，卻無法以金錢，甚至是食物引誘躺在窗邊的小貓咪，和牠交換到悠哉的一分鐘。

的確，每個人都可以用時間換取到金錢的報酬，但是可別忘了，就算我們的收入多到富可敵國，仍然無人可用任何的天價，去買到比其他人更多一點的時間。不僅如此，時間更是萬物的根源，因為有它，萬事萬物才有意義，才有生機；相反的，若沒它，所有一切都不存在了。所以，就算高明如科學家或哲學家能對浩瀚無際的太空侃侃而談，卻對時間的探究還是一知半解，因為時間是無法解釋的。

時間恰如其分地提供給我們需要，正如生命每天都會出現奇蹟一般。當我們想到這點時，就應該覺得它如神賜般的奇異恩典，充滿感激及驚喜。它讓每個人都各自擁有相同的時間，無異是生命中最彌足珍貴的寶藏。

當我們在清晨一睜開雙眼的剎那，就好像發現皮包裡，早已充塞著滿滿的時間可供使用，而且每天都是全新的，未被人動用的二十四小時。只要我們用心就可以發現，沒有人可以從我們手裡奪走任何的時間，更別說它會被偷去。在一天裡，沒有人可以擁有比任何人更多的時間，也不會更少。

在講求民主的時代，在時間的國度中，更可說是各國政府學習的典範。因為沒有其他事物，可以在它的領土中占有特權，即使是人人都夢寐以求的財富、智慧、健康……等。即便是天才，在一天中，也絕不可能比任何人能享有更多一點的時間。

除此之外，時間也是最公平而善良的主宰。不論我們如何毫不在乎地懶散荒廢，它既不會因而就停止供給，更不會為此而嚴懲我們。或許，若真有看不過去的天上神靈會說：「這個人如此浪費時間，若不是笨蛋就是傻子。既然他不值得擁有時間，還不如取消他的時間吧！」但實際上，我們卻始終沒有受過這種懲罰，我們的時間從來就未曾因為浪費，甚至是做錯任何事而遭受到中止，除了生命結束。

時間，亦如無可挑剔的最佳投顧公司。它的信用甚至比績優債券更好，連國定假日都全年無休地供給。它不會讓你成為負債累累的窮小子，最多只能讓你浪費過去的時間，即現在為止的以前的歲月；它也不會讓你預先領取未來的時間，就算是下一秒鐘，它也要完整地幫你保留到發生的當下，才肯支付。

因此有人說：「時間是生命中的奇蹟。」難道不是這樣嗎？

我們每天的生活都必須運用到二十四小時，不論是健康、快樂、財富、欲望和尊嚴，甚至包括心靈的淨化。生命中所有一切的事物，都必須仰賴時間才能完成。因此如何更積極有效地使用時間，對我們而言，是非常迫不及待要學習的事情。唯有學會正確地利用時間，掌控一天中的二十四小時，如此一來，我們渴望追求的幸福和理想，也才能真正地實現。

但奇怪的是，在現今事事講求快速和效率的時代裡，雖然經常可見報章傳媒披露關於「如何利用既有收入享樂生活」的文章，卻鮮見有人苦口婆心地宣傳「如何運用有限的時間樂活人生」的報導。

這個年代，金錢之所以變得比時間更重要，不僅是因為金錢無所不能的優勢，它可以滿足我們現實生活中各式各樣的需求，而且只要有意願賺取，生活中有許多方法都能幫助我們獲取更多的金錢，因此才會容易使人意識到其重要性及便利性。

但是一直在我們身邊默默付出的時間，卻有所不同。就在大部分的人充滿鬥志地努力向「錢」看時，它就毫無聲息地消逝在我們有限的生命中了。直到我們恍然大悟並悔恨不已的嘆息：原來，無論用多少錢都換取不回失去的青春。這一切都已經太遲了。

時間，對所有人都是規律而公平的。當我們在金錢上入不敷出時，會想盡各種方式彌補，不論是賺取、借貸、典當，或向社服機構求援，甚至有人不惜觸犯法律，都可以獲得想要滿足的數目。但不管我們用盡各種手段，每個人的一天就是二十四小時，無論你是總統或平民，天才或白痴都一樣。即使是老弱婦孺的時間，再強勢蠻橫的流氓想要如何巧取豪奪，它依然不為所動。

　　大環境不景氣的年代，物價又節節高升，一般薪水階級若只是靠每個月固有收入養家活口，許多家庭可能會感到苦不堪言。若想維持收支平衡，有些人選擇工作之餘再賺取更多的收入，而不想太辛苦的人，只要能節省生活開銷，安貧樂道一樣可以過活。總之再多麼困苦的日子，只要精打細算都可以熬過難關。

　　可惜的是，對使用金錢如此精明的現代人，卻沒有將同樣的態度運用到時間上。許多人在生活中，對金錢有著錙銖必較的靈敏度，對時間卻是迷迷糊糊地過了一生，也毫無知覺，如此一來豈不是本末倒置？時間雖然未曾停止過供應，但它還是極為精密地在計算著，一分一秒都不容我們忽視。

　　其實，財富只要用心努力就可以獲得，只是或多或少的數目問題。但一天二十四小時一旦恍恍惚惚地過了，便永遠失去它的「作用」，亦即，流失的可是我們寶貴的生命價值。更何況有多少人敢拍胸保證說：我在一天二十四小時中都過著充實的「生活」呢！

　　我所謂的「生活」，並非指單純的只是張口呼吸的「活著」，或者「糊里糊塗地過了一生」。重要的是我們用什麼態度及方式，在面對著經歷生命的過程。正如若有人為了自己付出一切努力，仍無法獲得大家認同，而感到懷憂喪志時，那他是否可化悲憤為力量，扭轉逆境？

　　或者當有人穿了一身筆挺的西裝，興高采烈地準備赴宴之時，他又是否能用心地注意身上的所有搭配細節，而不會因為某一樣小小的瑕疵，破壞了精心打扮的整體觀感？就如前例那位粗心的主人，在一心認為可以名貴器皿贏得賓客的所有目光之際，若能再更加留意盤中菜餚的滋味，是不是就更能達成賓主盡歡的宴客目的呢？那麼，包含我們自己在內的大部分人，又要到何時才能真正不再自欺欺人地說：「沒辦法啦！只要等我有多出來的時間，我一定會想辦法改掉它。」的這種壞習慣呢？

　　事實上大家心裡都很清楚，我們永遠不可能有「多出來的時間」。一直以來，我們所擁有的時間就是現在，當下的一切時間。至少我就從未感到，自己的時間能比別人充裕。

正因為如此，與我們人生息息相關的問題，就經常被人所忽略。所以，也促使了我想藉由大量的實地調查，以進一步瞭解一般人使用時間的方式，並提出關於有效的時間管理的十項重要策略，希望有助於大家重新省視自己，是如何看待這生命中最珍貴的資產——每一天中的二十四小時。

Chapter I

The Daily Miracle

"Yes, he's one of those men that don't know how to manage. Good situation. Regular income. Quite enough for luxuries as well as needs. Not really extravagant. And yet the fellow's always in difficulties. Somehow he gets nothing out of his money. Excellent flat--half empty! Always looks as if he'd had the brokers in. New suit-old hat! Magnificent necktie-baggy trousers! Asks you to dinner: cut glass-bad mutton, or Turkish coffee-cracked cup! He can't understand it. Explanation simply is that he fritters his income away. Wish I had the half of it! I'd show him--"

So we have most of us criticised, at one time or another, in our superior way.

We are nearly all chancellors of the exchequer: it is

the pride of the moment. Newspapers are full of articles explaining how to live on such-and-such a sum, and these articles provoke a correspondence whose violence proves the interest they excite. Recently, in a daily organ, a battle raged round the question whether a woman can exist nicely in the country on L85 a year. I have seen an essay, "How to live on eight shillings a week." But I have never seen an essay, "How to live on twenty-four hours a day." Yet it has been said that time is money. That proverb understates the case. Time is a great deal more than money. If you have time you can obtain money--usually. But though you have the wealth of a cloak-room attendant at the Carlton Hotel, you cannot buy yourself a minute more time than I have, or the cat by the fire has.

Philosophers have explained space. They have not explained time. It is the inexplicable raw material of everything. With it, all is possible; without it, nothing. The supply of time is truly a daily miracle, an affair genuinely astonishing when one examines it. You wake up in the morning, and lo! your purse is magically filled with twenty-four hours of the unmanufactured tissue of the universe of your life! It is yours. It is the most precious of possessions. A highly singular commodity, showered upon you in a manner as singular as the commodity itself!

For remark! No one can take it from you. It is unstealable. And no one receives either more or less than you receive.

Talk about an ideal democracy! In the realm of time there is no aristocracy of wealth, and no aristocracy of intellect. Genius is never rewarded by even an extra hour a day. And there is no punishment. Waste your infinitely precious commodity as much as you will, and the supply will never be withheld from you. Mo mysterious power will say:--"This man is a fool, if not a knave. He does not deserve time; he shall be cut off at the meter." It is more certain than consols, and payment of income is not affected by Sundays.

Moreover, you cannot draw on the future. Impossible to get into debt! You can only waste the passing moment. You cannot waste tomorrow; it is kept for you. You cannot waste the next hour; it is kept for you.

I said the affair was a miracle. Is it not?

You have to live on this twenty-four hours of daily time. Out of it you have to spin health, pleasure, money, content, respect, and the evolution of your immortal soul. Its right use, its most effective use, is a matter of the highest urgency and of the most thrilling actuality. All depends on that. Your happiness-the elusive prize that you are all clutching for, my friends!-depends on that. Strange that the newspapers, so enterprising and up-to-date as they are, are not full of "How to live on a given income of time," instead of "How to live on a given income of money"! Money is far commoner than time. When one reflects, one perceives that money is just about the commonest thing there is. It encumbers the earth in gross heaps.

If one can't contrive to live on a certain income of money, one earns a little more-or steals it, or advertises for it. One doesn't necessarily muddle one's life because one can't quite manage on a thousand pounds a year; one braces the muscles and makes it guineas, and balances the budget. But if one cannot arrange that an income of twenty-four hours a day shall exactly cover all proper items of expenditure, one does muddle one's life definitely. The supply of time, though gloriously regular, is cruelly restricted.

Which of us lives on twenty-four hours a day? And when I say "lives," I do not mean exists, nor "muddles through." Which of us is free from that uneasy feeling that the "great spending departments" of his daily life are not managed as they ought to be? Which of us is quite sure that his fine suit is not surmounted by a shameful hat, or that in attending to the crockery he has forgotten the quality of the food? Which of us is not saying to himself--which of us has not been saying to himself all his life: "I shall alter that when

I have a little more time"?

We never shall have any more time. We have, and we have always had, all the time there is. It is the realisation of this profound and neglected truth (which, by the way, I have not discovered) that has led me to the minute practical examination of daily time-expenditure.

第二章
策略一：欲望是動力而非結果

永無止盡的欲望或需求，其實也是生命中的一部分。

但是問題是這麼多的渴望，卻想要在這麼少的時間裡
完成，

這其中必然會在生命中發生許多的衝撞及掙扎。

每當我在公開場合討論到每個人每天時間的使用方式時，總會遇到某些人以不屑的口吻說：

「但是……喔！怎麼會有人不知道要如何善用二十四小時呢？像我呢，一向就不會有時間管理上的困擾。一天的時間，不但可以完成許多事情，甚至還有許多空閒的時間，還常用來參加媒體舉辦的娛樂競賽活動呢！其實，只要能明白清楚地認知，每一個人，每天都只有二十四小時，自然就會妥善地規劃生活，以滿足自己的需求。安排時間，這有什麼困難的嘛！」

於是緊接著，就會有人不以為然地回應：

「事實真是如此嗎？親愛的朋友，我長年以來尋尋覓覓，能幫我解決關於時間不足問題的人，原來就是你啊！那麼，請你告訴我，到底要付出多大的代價，你才願意告訴我這其中的祕訣。至今都還沒能向你請教，這可是我人生中的一大憾事。你可知道，在你尚未出現之前，包含我的許多人，每天都只能活在壓力和後悔之中。因為，我們只能眼睜睜地看著時間一天一分一秒地流逝，生活卻依然充滿許多無奈的痛苦。」

不可否認，當我們遇見有人自吹自擂地說，自己是如何地多麼地會計畫生活時，我們可能都曾在剎那間，產生上述說話者的心態。但是，倘若用心分析一下這種感覺，將不難發覺這種想法，正反應出我們對於歲月所產生的渴望及焦慮。

因為在生命中，永遠有無止盡的欲望等著要填滿，但是問題是，有限的歲月又要如何來得及滿足我們的需求呢？

因此時間對生活所造成的壓迫感，便經常使我們的生活籠罩在一片陰影之中。

就猶如在值得慶賀的歡樂場合中，不時出現一個令大家避之唯恐不及的討厭鬼（時間），非但讓人都快樂不起來，滿腦子就只想逃離現場。又如，當我們正在戲院樂不可支地觀賞一部喜劇片時，那個像背後靈般的討厭鬼，不知不覺地又倏忽出現在我們眼前，而且它還在瞬間化為一根教鞭，直挺挺地指著我們嘮叨著。

　　或者在我們玩樂一整天後，正要拚命飛奔趕搭最後一班電車，就在好不容易抵達月臺時，那個如影隨形的討厭鬼不僅跟在我們身邊，還冷冷地睨著我們：「嘿，朋友，你到底用你寶貴的時間做了哪些事？看看你，一天到晚匆匆忙忙地跑來跑去，但到底都在忙些什麼『有意義的事』呢？」

　　生命的價值最終在於實現自我，過程中必然會產生許多需求及渴望。然後，透過一個又一個階段的欲望滿足或期待實現，才能完成我們生命的任務。

　　有人說：「永無止盡的欲望或需求，其實也是生命中的一部分。」而事實也的確是如此。但是，問題是這麼多的渴望，卻想要在這麼少的時間裡完成，這其中，必然會在生命中發生許多的衝撞及掙扎，當然，也就可能因而混淆了生命初衷的焦點。

　　因此，我們首先必須要確實地瞭解欲望的動機及可能性。就如有一個人想要給自己一個悠閒的長假，一開始，他只是單純地聽到內心的一股聲音揚起：「去吧！去旅行吧！不要再猶豫了。」於是他決定將計畫實現。

接下來他開始盤算著：那麼，究竟是參加旅行社精心規劃的行程比較安全呢？還是一手策劃自助旅行，比較有趣呢？緊接著，他算算經費後發現：喔，好像只能去亞洲國家玩玩耶！可是又好想去歐洲和美洲，怎麼辦呢？需要借貸嗎？不幸的是，他又想到更多的問題了：雖然想去海邊玩，但會不會因為不會游泳而溺斃在海裡，那是否要先去學游泳？唉！還是多買些保險？喔！意外險是一定要買的，如果出車禍受傷了就有補助。

層出不窮的問題顯然讓他開始有些動搖。

除此之外，近年來旅行意外的頻傳，尤其是天災更是難以預料，於是他又想到：若是到了印尼又遇到地震或海嘯時，他又要準備哪些救災品呢？還有如何防範？或是避免去哪些國家呢？最後他終於下了另一個決定：一天只有二十四小時，天啊！我的時間哪來得及解決這麼多問題，還是別去吧！恐懼就這樣終結了他最初的夢想。

更不幸的事還不僅於此，因為最後不論他有沒有實現出國旅遊的欲望，去或不去都成為他在往後長久的日子裡，一個永遠擺脫不掉的夢魘。結果，與那些從未想到要去環遊世界的人們相比，他寧可自己從未產生過這樣的想法。因為，至少他們現在不會比他更煩惱的了。

其實對大多數的人來說，旅行只是釋放壓力的一種方式。如果只是單純地想要放鬆身心，即使就近來趟國內環島的自助旅行，也是非常愜意的。何況還有許多人，可能這一生中都尚未有機會搭乘飛機出國呢！上例中的主角雖然因為太多的問題，而放棄了環遊世界的夢想，他歸咎於沒有太多的時間去解決問題。但他真正最大的問題，卻是他未曾想到要善用時間克服困難，他甚至連到旅行社去詢問相關資訊的機會都沒有給自己，就更別說想用時間解決其他無關緊要的事了。

總之不論是夢想或是計畫，最重要的還是要落實地去執行。每個人都清楚地知道，每天就只有二十四小時，沒有人有例外。即使忙碌的程度有所不同，但是只要有意願，支配時間的掌控權卻是操之在我們自己的手上。

就好像有人一天到晚叫嚷著要跟團去旅行，朋友也熱心地給了旅行社的聯絡資料，結果過了大半個月還是毫無動靜。因為他的理由是：唉，我實在忙到連打電話的時間都沒有啦！他終於說出問題癥結——他連打電話給旅行社的意願都沒有。

若進一步來分析，為何我們會產生這些不堅定而又蠢蠢欲動的想法，可以發現它其實來自於：一種長久以來積非成是的想法。因為我們總是以為，除了完成現階段自己應該做的事之外，我們有能力可以讓自己及家人，過著再更加優渥的生活，所以便產生了新的欲望。

因此不管是為了法律上的規定、責任心、企圖心和社會輿論，於是我們除了想擁有健全快樂的家庭，還想供給親人富裕無虞的生活，此外還要滿足工作及興趣的種種需求，所以我們必須要更努力地賺錢，用心地儲蓄，並積極地償還負債，而且還要學習透過適當正確的理財投資，以增加更多的財富。

但是當我們全力以赴朝向目標前進時，就會發現這項任務其實一點也並不容易。事實上，不僅沒有人可以盡善盡美地完成，而且過程中的許多情況經常是事與願違，結果總讓人難以預料。雖然最後，依然有人可以幸運擁有幸福而富裕的家庭，但卻永遠會為了更多新的需求，而始終對生活感到不滿足。於是不斷增生的欲望便如魍魎一般，無時無刻地纏繞著我們。

即使到最後有人終於瞭解了，有些事非窮自己一生的能力所能及，有些渴望也不是全力以赴就會達成。但還是有人認為，只要多付出一分心力，就可以降低一分失望。就算到了精疲力盡的時刻，還是要為夢想中的欲望拚戰到最後一刻，這種精神著實令人可敬可佩。尤其對那些不斷攀登高峰的成功者而言，他們的人生原本就是為了滿足無止盡的欲望而生存。

但是同時我們也都瞭解，為了要實現欲望，往往也要付出某種程度甚至更高的代價。這其中當然包括在欲望滿足之前，那種因為想要有所行動，卻又不能行動的焦慮感。它會一直不斷地困擾著我們原本平靜的心靈。諸如這樣的情緒及諸多不同的欲望，心理學家們也都已經賦予了各種形式的定義，就如同我們熟知的求知欲，也是欲望其中的一種形式。

　　至於求知欲，究竟可以有多強烈呢？看看那些為了建構出系統性知識，而奉獻一生心力的專家吧！他們的付出，甚至可以超越自我的極限，就如我最敬仰的思想家赫伯特‧斯賓塞（Herbert Spencer，註1），也不能克制住這種需求無度的求知欲。因此，據說他經常會將自己陷入研究論題中不斷地思索，不斷地疑惑，以至於到無法自拔的地步。

　　在社會上，不乏有許多充滿智慧並具有求知欲的人。他們不但意識到自己生存的目的，同時也瞭解自己對生命所擁有的欲望。當他們產生想要超越現有目標的渴求時，經常會藉由閱讀以達成目標。其中，又有許多人是極為熱愛文學的。

　　但是要提醒讀者的是，並非只有文字的表現，才能涵蓋所有知識。要想提升自我的知識水準並滿足求知欲，除了書籍以外，其實還有許多的方式，這便是在之後的章節所將闡述的內容。

一天始於下班後：
妥善運用你的待機時間

註 1 赫伯特‧斯賓塞（Herbert Spencer，1820 － 1903 年），英國哲學家，著有《社會學原理》，為闡明社會學分析的系統著作。他的社會學中將家庭的發展居於首位，並研究社會各部分的相互關係。

Chapter II

The Desire to Exceed One's Programme

" But," someone may remark, with the English disregard of everything except the point, "what is he driving at with his twenty-four hours a day? I have no difficulty in living on twenty-four hours a day. I do all that I want to do, and still find time to go in for newspaper competitions. Surely it is a simple affair, knowing that one has only twenty-four hours a day, to content one's self with twenty-four hours a day!"

To you, my dear sir, I present my excuses and apologies. You are precisely the man that I have been wishing to meet for about forty years. Will you kindly send me your name and address, and state your charge for telling me how you do it? Instead of me talking to you, you ought to be talking to me.

Please come forward. That you exist, I am convinced, and that I have not yet encountered you is my loss. Meanwhile, until you appear, I will continue to chat with my companions in distress--that innumerable band of souls who are haunted, more or less painfully, by the feeling that the years slip by, and slip by, and slip by, and that they have not yet been able to get their lives into proper working order.

If we analyse that feeling, we shall perceive it to be, primarily, one of uneasiness, of expectation, of looking forward, of aspiration. It is a source of constant discomfort, for it behaves like a skeleton at the feast of all our enjoyments. We go to the theatre and laugh; but between the acts it raises a skinny finger at us. We rush violently for the last train, and while we are cooling a long age on the platform waiting for the last train, it promenades its bones up and down by our side and inquires: "O man, what hast thou done with thy youth? What art thou doing with thine age?" You may urge that this feeling of continuous looking forward, of aspiration, is part of life itself, and inseparable from life itself. True!

But there are degrees. A man may desire to go to Mecca. His conscience tells him that he ought to go to Mecca. He fares forth, either by the aid of Cook's, or unassisted; he may probably never reach Mecca; he may drown before he gets to Port Said; he may perish ingloriously on the coast of the Red Sea; his desire may remain eternally frustrate. Unfulfilled aspiration may always trouble him. But he will not be tormented in the same way as the man who, desiring to reach Mecca, and harried by the desire to reach Mecca, never leaves Brixton.

It is something to have left Brixton. Most of us have not left Brixton. We have not even taken a cab to Ludgate Circus and inquired from Cook's the price of a conducted tour. And our excuse to ourselves is that there are only twenty-four hours in the day.

If we further analyse our vague, uneasy aspiration, we

shall, I think, see that it springs from a fixed idea that we ought to do something in addition to those things which we are loyally and morally obliged to do. We are obliged, by various codes written and unwritten, to maintain ourselves and our families (if any) in health and comfort, to pay our debts, to save, to increase our prosperity by increasing our efficiency. A task sufficiently difficult! A task which very few of us achieve! A task often beyond our skill! yet, if we succeed in it, as we sometimes do, we are not satisfied; the skeleton is still with us.

And even when we realise tat the task is beyond our skill, that our powers cannot cope with it, we feel that we should be less discontented if we gave to our powers, already overtaxed, something still further to do.

And such is, indeed, the fact. The wish to accomplish something outside their formal programme is common to all men who in the course of evolution have risen past a certain level.

Until an effort is made to satisfy that wish, the sense of uneasy waiting for something to start which has not started will remain to disturb the peace of the soul. That wish has been called by many names. It is one form of the universal desire for knowledge. And it is so strong that men whose whole lives have been given to the systematic acquirement of knowledge have been driven by it to overstep the limits of their programme in search of still more knowledge. Even Herbert Spencer, in my opinion the greatest mind that ever lived, was often forced by it into agreeable little backwaters of inquiry.

I imagine that in the majority of people who are conscious of the wish to live--that is to say, people who have intellectual curiosity--the aspiration to exceed formal programmes takes a literary shape. They would like to embark on a course of reading. Decidedly the British people are becoming more and more literary. But I would point out that literature by no means comprises the whole field of knowledge, and that the disturbing thirst to improve one's self--to increase one's knowledge--may well be slaked quite apart from literature. With the various ways of slaking I shall deal later. Here I merely point out to those who have no natural sympathy with literature that literature is not the only well.

第三章
策略二：執行前的心理調適

世界上找不到任何一種方法，
可以真正地教我們如何「開始」過日子的。
親愛的朋友，那就「開始」生活啊！
老老實實地使用你一天中的二十四個小時。

假若我先前的觀點，能取得你們初步的認同，相信
你們應該是正為了「時不我與」而感到生命中飽受挫折，
所以才能如此深刻地體會到時間的「重要性」。

每個人，每天都會有許多想做而尚未完成的事。而
我們也總是幻想著，要是能再有「多一點的時間」該多
好啊！那必定可以成就更多自己的想望。但事實的真相
是：永遠都不可能有「多一點的時間」這種事發生。要
知道，我們所擁有的時間，就是當下所有的一切。

因此有人或許會引頸期盼能學到關於「如何善用一
天二十四小時」的這種訣竅，好讓自己能揮別整年來因
為一事無成而籠罩的滿天烏雲。但很遺憾的，直到現在
就連我自己，也未曾發現過這樣的好消息。但就算真的
有，我也不想發現，更不希望有人能找到，因為那終究
只是不切實際的幻想。正如你想前往麥加朝聖，雖然遙
遠的路途中必定險象環生，但更悲慘的是你卻根本沒有
圓夢的機會。

在你能確切體認時間的本質前，或許會因為那個不
知從何而來的「祕訣」，而喚起新的夢想。你會樂不可

支地告訴自己：「太棒了！我將學到一種不費吹灰之力的時間管理法，從此就可以輕而易舉地完成許多想做的事。」事實是，這麼一來，你將離真相越來越遙遠了。因為時間的運用，根本沒有所謂簡單的或困難的方法，而且你只有一種唯一的方法，那就是：老老實實地使用你一天中的二十四個小時。

當我們面對生命時，計畫是永遠擋不住變化的。我們不能預估何時會遭逢怎樣的考驗，一旦有了投機取巧的念頭，做起事來就會眼高手低且不切實際，如此，幸福人生終究只是遙不可及的白日夢！要知道，若想要能充實而快樂的度過所謂「有意義的每一天」，最重要的是要能瞭解，在二十四小時中有其生活上的高難度，亦即它所要付出全力以赴的心血及堅持不懈的毅力，而你又能做到哪種程度呢？

不要天真地以為只要認真地坐在書桌前，然後寫下一張羅列詳細的計畫表，就是完成了人生的規劃。因為若是面對未來可能發生的挫折及失敗，你都還未能做好心理準備迎戰，那麼當照著計畫表進行，事情一旦生變時，只會讓你手忙腳亂得不知如何應變。

又或者你是個相信有所付出，就一定會有所收穫的人。如此一來，當計畫因故取消，或是花盡心血卻只換來蠅頭小利時，當你看著再也不能更精確的計畫表，你可能不只是捶胸頓足、扼腕嘆息，甚至心裡還後悔，早知如此，還不如當個只會睡覺和呼吸的「普通人」。

其實人生本就無常，順境與逆境，往往就在一瞬之間形成。不想或不能應變的人，有人說：那就只要認命地「活著」就好了。但是生命的價值若只是「活著」，這樣的人生未免也太不幸了吧！但是我倒認為，一個人若真懂得真實地活在當下，就算他只會張口呼吸地「生存」著，他的生命也有了意義。

我們要想經歷充實而精采的生命旅程，本來就必須學會真實地感受到「存在」的意義及其重要性。生活是需要學習智慧並充滿勇氣的，當我們充滿信心時，才能接受人生中無可預料的各種挑戰。所以說：「我思故我在。」不就是因為人類懂得去思考生命，實踐人生，才能比較出和其他生物間的差異性嗎？

　　因此你說：「是的，我認同前述的話，也願意接受你的建議。面對人生，我已決定全力以赴地努力生活。可是，又要從何開始呢？」

　　親愛的朋友，那就「開始」生活啊！世界上找不到任何一種方法，可以真正地教我們如何「開始」過日子的。就好比當你站在泳池邊，有一個正想要玩水的人問你：「請問，我要怎樣才能到池裡去坑呢？」我們只要告訴他：「別害怕，你只要直接下水就可以了，不管用什麼方法。」

　　對於源源不絕的時間來說，它最公平之處，就在於不會讓誰預支先用。不管是明年、明天、下一個小時，它都是完整封存、未被使用的，彷彿我們未曾浪費過任何時間。這一點是否讓我們感到備受恩寵呢？彷彿在一生中，永遠都有從天而降的預存儲糧。因此只要我們想有所作為，人生便能隨即重新開始。

　　當下，想做就做！

　　別再等到下一個星期或明天，那都是毫無意義的藉口。就如同一個想要下池玩水的人，但一碰到冰涼的水卻又將腳縮回，他想著：「還是下次水暖和些再來玩吧！」奉勸你，就別想太多了。就是現在，下水玩吧！因為事實上，水不但可能變得更冷，而你可能也不會想再來了。但是，在你開始進行之前，有兩項原則還是需要留意的：

第一、請維持平常心。

生命雖然需要適度的熱情來加溫，但不要因一時激情的影響而衝動行事，以免樂極生悲，遭到無法挽回的下場。

因為若是一開始就表現出過度的熱情，一味地往前衝、衝、衝的結果，不但會有誤闖陷阱的危機，若是到了最後表現差強人意時，你更會因此感到不甘心，而永遠無法滿足現狀。於是你為了能夠得到更多，便會不惜一切地付出更多，雖然是不顧一切地勇往直前，甚至到超出自己能力所及的地步，最後非但無法保證就能一償夢想，說不定還會因此得不償失。

更何況，當事情不斷發生挫敗時，最初旺盛的鬥志也會逐漸因為當初自以為是的蠻幹而消耗殆盡，此時你除了只能自我安慰：「夠了，你已經盡力了。」從此，你很可能再也沒有勇氣接受人生的任何挑戰。

第二、勿輕視任何小小成就的影響。

任何事都不可能在一開始就順暢進行，因此即便是小小的收穫，都要懂得感恩地收藏起成果，留存在奮鬥的路上，不斷地作為鼓勵打氣之用。

更何況世事難料，我們除了要盡其所能地盡人事外，特別是要懂得克服人類的劣根性，之後也才能安心自在地聽天命。要知道，大部分失敗的人生，都是因為貪心不足所致。

在一天短暫而有限的時間裡，確切地朝著夢想的生活落實而做，無須要求自己每天都要成就太多、太高的希望，但是也要避免落入太早或過多的挫敗之中，以免造成失敗的連鎖效應。

當然一次兩次的挫折在所難免，要相信自己能因失敗而更成長，並強化自尊心而不會輕易再被擊倒。基本上，我並不認同所謂「光榮的失敗勝過微不足道的成功」的論調。相反的，我認為即使再渺小的成果，都可以為生活帶來更多的喜悅，為生命增添更美的顏色。反之，就算失敗的原因是多麼讓人感到悲痛，終究還是不幸的「一場空」吧！對時間的利用價值來說，它的成績就是一無所獲。

有許多人認為自己在安排時間上，可說已經到了盡心盡力的地步，因為滿滿的時間表上，根本找不到一點浪費的痕跡。但真相果真是如此嗎？若是扣除掉工作的七個小時（或八個小時），及睡眠的七個小時，如果想讓自己有充足的休息，其實睡上九個小時也無妨。計算下來，你一天的二十四個小時還有八個小時可使用，這些時間你究竟是如何支配的呢？同時，我們也需要做更進一步的檢視。

下一章我們就將深入地來探討，在支配時間時經常會出現的各種問題，如：何謂充分地使用時間、工作所占用的部分、睡眠時間、剩餘時間的利用……等。

一天始於下班後：
妥善運用你的待機時間

Chapter III
Precautions Before Beginning

Now that I have succeeded (if succeeded I have) in persuading you to admit to yourself that you are constantly haunted by a suppressed dissatisfaction with your own arrangement of your daily life; and that the primal cause of that inconvenient dissatisfaction is the feeling that you are every day leaving undone something which you would like to do, and which, indeed, you are always hoping to do when you have "more time"; and now that I have drawn your attention to the glaring, dazzling truth that you never will have "more time," since you already have all the time there is--you expect me to let you into some wonderful secret by which you may at any rate approach the ideal of a perfect arrangement of the day, and by which, therefore, that haunting, unpleasant, daily disappointment of things left undone will be got rid of!

I have found no such wonderful secret. Nor do I expect to find it, nor do I expect that anyone else will ever find it. It is undiscovered. When you first began to gather my drift, perhaps there was a resurrection of hope in your breast. Perhaps you said to yourself, "This man will show me an easy, unfatiguing way of doing what I have so long in vain wished to do." Alas, no! The fact is that there is no easy way, no royal road. The path to Mecca is extremely hard and stony, and the worst of it is that you never quite get there after all.

The most important preliminary to the task of arranging one's life so that one may live fully and comfortably within one's daily budget of twenty-four hours is the calm realisation of the extreme difficulty of the task, of the sacrifices and the endless effort which it demands. I cannot too strongly insist on this.

If you imagine that you will be able to achieve your ideal by ingeniously planning out a time-table with a pen on a piece of paper, you had better give up hope at once. If you are not prepared for discouragements and disillusions; if you will not be content with a small result for a big effort, then do not begin. Lie down again and resume the uneasy doze which you call your existence.

It is very sad, is it not, very depressing and sombre? And yet I think it is rather fine, too, this necessity for the tense bracing of the will before anything worth doing can be done. I rather like it myself. I feel it to be the chief thing that differentiates me from the cat by the fire.

"Well," you say, "assume that I am braced for the battle. Assume that I have carefully weighed and comprehended your ponderous remarks; how do I begin?" Dear sir, you simply begin. There is no magic method of beginning. If a man standing on the edge of a swimming-bath and wanting to jump into the cold water should ask you, "How do I begin to jump?" you would merely reply, "Just jump. Take hold of your nerves, and jump."

As I have previously said, the chief beauty about the constant supply of time is that you cannot waste it in advance. The next year, the next day, the next hour are lying ready for you, as perfect, as unspoilt, as if you had never wasted or misapplied a single moment in all your career. Which fact is very gratifying and reassuring. You can turn over a new leaf every hour if you choose. Therefore no object is served in waiting till next week, or even until to-morrow. You may fancy that the water will be warmer next week. It won't. It will be colder.

But before you begin, let me murmur a few words of warning in your private ear.

Let me principally warn you against your own ardour. Ardour in well-doing is a misleading and a treacherous thing. It cries out loudly for employment; you can't satisfy it at first; it wants more and more; it is eager to move mountains and divert the course of rivers. It isn't content till it perspires.

And then, too often, when it feels the perspiration on its brow, it wearies all of a sudden and dies, without even putting itself to the trouble of saying, "I've had enough of this."

Beware of undertaking too much at the start. Be content with quite a little.

Allow for accidents. Allow for human nature, especially your own.

A failure or so, in itself, would not matter, if it did not incur a loss of self-esteem and of self-confidence. But just as nothing succeeds like success, so nothing fails like failure. Most people who are ruined are ruined by attempting too much. Therefore, in setting out on the immense enterprise of living fully and comfortably within the narrow limits of twenty-four hours a day, let us avoid at any cost the risk of an early failure. I will not agree that, in this business at any rate, a glorious failure is better than a petty success. I am all for the petty success. A glorious failure leads to nothing; a petty success may lead to a success that is not petty.

So let us begin to examine the budget of the day's time. You say your day is already full to overflowing. How? You actually spend in earning your livelihood--how much? Seven hours, on the average? And in actual sleep, seven? I will add two hours, and be generous. And I will defy you to account to me on the spur of the moment for the other eight hours.

第四章
策略三：時間分配的原則

很多人的「通病」就是，只將工作的早上十點到晚間
六點視為有意義的「一天」。
至於早上十點以前的時間，及晚間六點以後的時間，
都只是這「一天」的開始及結束。

　　為了幫助讀者能在最短期間內確實瞭解善用時間的原則，我將用一名上班族約翰作為實例解說。其中要跟讀者說明，由於每個人都具有其性格習性，因此在使用時間上，就會產生差異性的存在。關於約翰所經歷的事，雖未能完全代表大家必然會發生，但我們仍可就其普遍存在的慣性來作為參考的依據。

　　約翰的工作時間是從上午十點到晚間六點為止。每一天，早晚他都要花費將近各五十分鐘的時間在交通上。從時間的角度來看，不管約翰和他的老闆的收入差距十萬八千里，但是他們在公司所擁有的時間，是一樣的多。

　　但是在面對時間的態度上，使得約翰和別人的使用時間開始產生差別了，當然也由於這個錯誤，使他有三分之二的心力及生活樂趣，就這樣在生活中消失殆盡。事實上，約翰對自己的工作已經感到索然無味，至少在許多時候都是這樣的。簡單地說，是只差一步就到厭惡的地步了。

　　讓我們先來看看他上班的狀況吧！每天一到開始工

作的時刻，他總是心不甘情不願地坐到位子上，而且是盡可能地使出各種拖延戰術。但到了下班時間，他可就判若兩人了，通常他都是以迅雷不及掩耳的速度，興奮而快步地離開公司。若綜觀他整天的工作績效，不但乏善可陳，簡直就連例行性事務也沒有處理好。以老闆的立場來說，約翰的行為及效率，真的已經到了「米蟲」的地步。

　　或許有讀者會認為，我這樣描述過於嚴苛，但我只是說出真相。難道各位從未曾在公司裡看過這樣的同事嗎？對他們的行徑，或許你心裡早就不以為然了，更何況真正要注意的是，請別讓自己成為他人心目中的「米蟲」！

　　除此之外，另一個問題是很多人的「通病」。那就是，只將工作的早上十點到晚間六點（占一天中的三分之一）才視為有意義的「一天」。至於早上十點以前的時間，及晚間六點以後的時間，都只是這「一天」的開始及結束。

　　當然這種感覺其實都是習慣養成，但不知不覺中這

「一天」以外的十六個小時（一天中的三分之二），似乎都成了我們多餘的時間。就算在十六個小時裡不至於讓光陰虛度，但我們卻很容易地就忽略了它們的存在。另外，如果在一般人的心裡，認為「一天」中至少要能完成「應該要做的工作」，才算是度過有意義的「一天」。果真如此，那約翰就和許多人一樣，更難自圓其說了。

因為他們若將那三分之二的時間，視為「一天」中三分之一以外多餘的時間，他們就應該要對那真正而重要的三分之一的時間，亦即對「應該要做的工作」用心地全力以赴才對。但事實卻非如此，真相是他們在工作時，大多數的情況還是「心不在焉」地虛晃一天。那就更別說，他們會如何地妥善利用那剩餘的十六小時？

約翰若是真心期望自己不要再浪費生命，完整而充實地度過「一天」，他應該要學會在工作之餘的「一天」（上午十點到晚間六點），再給予自己另外的「一天」（晚間六點後至上午十點前）。

這一天，他將擁有十六個小時，完完全全屬於自己的時間。這時，他將是自由之身，不需要再為生計煩惱。他可以將這十六小時用來使自己更健康、培養人際關係、鍛鍊專長、多和家人共處……亦即，多做一些對自己有益的事。

總之，只要能妥善地利用這十六小時，未來他所獲得的一切，都會比每天幻想著從天而降的遺產還要來得更多。對於這原先以為是多餘的「一天」（下班後的十六小時），只要他願意改變態度，他的人生也會隨之有所變化。

無須擔心，即使你使出渾身解數度過十六個小時的生活，也不會因此就讓大腦發生過勞死，甚至影響到工作時的八個小時的效率。相反的，一個對生命充滿好奇及興趣的你，只會更加活力十足地提高工作效率。要知道，我們的大腦除了需要適當而充足的睡眠外，它可是經常充滿幹勁的，隨時準備接受事物不斷的刺激變化。因此，不要以為它會像四肢一樣，常常會感到肌肉痠痛「疲勞」。

接著讓我們再進一步來觀察，約翰在這十六個小時中的狀況，來瞭解他是如何安排自己的生活。該注意的是我們並非在意他支配時間的長短，重點是在於他做的事是否為必要性。正如一個拓荒新手想要開闢新地，教他瞭解如何開拓比知道何時去開拓是更為重要的。

　　一般來說，約翰會在九點準時起床，接著在九點七分到九分間用完早餐；通常他都會在九點十分前，就鎖門外出。截至目前，他似乎還沒有浪費到時間。但問題就在他關上大門之後，頓時，他覺得自己像個機器人般的空洞，雖然腦袋空白一片，卻又可以本能地走向站牌，然後像個傻子一樣等著，等著，直到車子來了。

　　但是這段等車的時間，也就這樣無聲無息地消失了。每天單是這段時間，在數不清的公車站、捷運車站……我們都可以看到許多人目光呆滯地等車、上車和下車。如果把大家的這段時間核計一下，並換算成金錢，恐怕比當日股市流動的資金量要多上許多吧！但我們卻在哈欠、瞌睡和發呆中，把時間給遺棄了。

　　換另一種方式來看，我們將這十六個小時，等同於一張可花用的時間紙鈔吧。這張紙鈔，也可以兌換成等值的不同幣值的時間硬幣。在使用的過程中，我們經常會漫不經心地就浪費掉幾個時間硬幣。

　　但是倘若有人告訴我們：「你若是要兌換時間紙鈔或任何硬幣，必須要再加收三十元手續費。」這時，通常每個人都會恢復神志，然後開始精明地計算起來，甚至還強悍地討價還價。顯然我們都忘記了，剛剛等車的時候，一個又一個的閃亮亮時間硬幣，不也是在我們眼前憑空消失嗎？這與我們平日在使用時間的方式，何嘗不是一樣呢！

　　總而言之，別小看零零碎碎的時間。所謂積沙成塔，只要能善加利用時間，日子一久，這些微不足道的「時間硬幣」，對我們生命所產生的影響是不容小覷的。

Chapter IV

The Cause of the Troubles

In order to come to grips at once with the question of time-expenditure in all its actuality, I must choose an individual case for examination. I can only deal with one case, and that case cannot be the average case, because there is no such case as the average case, just as there is no such man as the average man. Every man and every man's case is special.

But if I take the case of a Londoner who works in an office, whose office hours are from ten to six, and who spends fifty minutes morning and night in travelling between his house door and his office door, I shall have got as near to the average as facts permit. There are men who have to work longer for a living, but there are others who do not have to work so long.

Fortunately the financial side of existence does not interest us here; for our present purpose the clerk at a pound a week is exactly as well off as the millionaire in Carlton House-terrace.

Now the great and profound mistake which my typical man makes in regard to his day is a mistake of general attitude, a mistake which vitiates and weakens two-thirds of his energies and interests. In the majority of instances he does not precisely feel a passion for his business; at best he does not dislike it. He begins his business functions with reluctance, as late as he can, and he ends them with joy, as early as he can. And his engines while he is engaged in his business are seldom at their full "h.p." (I know that I shall be accused by angry readers of traducing the city worker; but I am pretty thoroughly acquainted with the City, and I stick to what I say.)

Yet in spite of all this he persists in looking upon those

hours from ten to six as "the day," to which the ten hours preceding them and the six hours following them are nothing but a prologue and epilogue. Such an attitude, unconscious though it be, of course kills his interest in the odd sixteen hours, with the result that, even if he does not waste them, he does not count them; he regards them simply as margin.

This general attitude is utterly illogical and unhealthy, since it formally gives the central prominence to a patch of time and a bunch of activities which the man's one idea is to "get through" and have "done with." If a man makes two-thirds of his existence subservient to one-third, for which admittedly he has no absolutely feverish zest, how can he hope to live fully and completely? He cannot.

If my typical man wishes to live fully and completely he must, in his mind, arrange a day within a day. And this inner day, a Chinese box in a larger Chinese box, must begin at 6 p.m. and end at 10 a.m. It is a day of sixteen hours; and during all these sixteen hours he has nothing whatever to do but cultivate his body and his soul and his fellow men. During those sixteen hours he is free; he is not a wage-earner; he is not preoccupied with monetary cares; he is just as good as a man with a private income. This must be his attitude. And his attitude is all important. His success in life (much more important than the amount of estate upon what his executors will have to pay estate duty) depends on it.

What? You say that full energy given to those sixteen hours will lessen the value of the business eight? Not so. On the contrary, it will assuredly increase the value of the business eight. One of the chief things which my typical man has to learn is that the mental faculties are capable of a continuous hard activity; they do not tire like an arm or a leg. All they want is change--not rest, except in sleep.

I shall now examine the typical man's current method of employing the sixteen hours that are entirely his, beginning with his uprising. I will merely indicate things which he does and which I think he ought not to do, postponing my suggestions for "planting" the times which I shall have cleared--as a settler clears spaces in a forest.

In justice to him I must say that he wastes very little time before he leaves the house in the morning at 9.10. In too many houses he gets up at nine, breakfasts between 9.7 and 9.9 1/2, and then bolts. But immediately he bangs the front door his mental faculties, which are tireless, become idle. He walks to the station in a condition of mental coma. Arrived there, he usually has to wait for the train. On hundreds of suburban stations every morning you see men calmly strolling up and down platforms while railway companies unblushingly rob them of time, which is more than money. Hundreds of thousands of hours are thus lost every day simply because my typical man thinks so little of time that it has never occurred to him to take quite easy precautions against the risk of its loss.

He has a solid coin of time to spend every day--call it a sovereign. He must get change for it, and in getting change he is content to lose heavily.

Supposing that in selling him a ticket the company said, "We will change you a sovereign, but we shall charge you three halfpence for doing so," what would my typical man exclaim? Yet that is the equivalent of what the company does when it robs him of five minutes twice a day.

You say I am dealing with minutiae. I am. And later on I will justify myself. Now will you kindly buy your paper and step into the train?

第五章
策略四：創造時間的高價值

請不要忘記，我們都不是時間的主人。
任何人所擁有的時間是一樣多的。
盡量減少在清晨的車廂裡看報吧！
就在同時，有許多人已經在利用時間，
將更多有意義的時光存進他的生命帳戶裡。

終於搭上早晨的電車了。你知道自己足足有三十分鐘的「多餘時間」，於是不假思索地拿起水果日報，準備悠閒地閱覽一番。

你的視線從一則又一則的新聞往下看，就連滿頁的廣告也未曾放過。當你的焦點從旅遊及影視八卦的版面移動時，這一幅輕鬆自在的畫面，你彷彿是遠離俗世的修行者。就在這屬於自己的時間國度裡，你的這一天，好像已經從二十四小時延伸為一百二十四小時。

我也喜歡閱讀報紙，每天必看五份英文報及兩份法文報。但我想可能還是只有報攤老闆會比較清楚我實際看報的數量及次數吧！事實上，在我每天的行程計畫中，並沒有閱報這一項。通常，我只利用極為零碎及空閒的時間來看報，但重要的是我會非常認真地詳讀。

閱報，純然是個人的興趣。基本上，我不認為在早晨看報是適宜的。這並非代表我對報紙有任何偏見，而是我認為報紙在匆忙間付印，在急促間被閱讀，錯誤的訊息也可能在不經意間就在腦海裡形成了，甚至會影響一天工作的心情。

　　但最可惜的，還是有人用一大清早三四十分鐘的時間來看報，這才最浪費了珍貴的時間。就像一個浪費奢侈的暴發戶，隨意地揮霍他得來容易的財富，將如鑽石般寶貴的清晨用來看報，尤其是八卦新聞，更是對時間的暴殄天物。

　　所謂「一日之計在於晨」，何不在或許有些擁擠的車廂中，安靜地沉思於自我之中，這樣或許對你的人生更有意義。

　　請不要忘記，我們都不是時間的主人。更何況，任何人所擁有的時間是一樣多的。為了使生命能發揮更多更大的潛能，盡量減少在清晨的車廂裡看報吧！**就在你閱報的同時，有許多人已經在利用時間，將更多有意義的時光存進他的生命帳戶裡。**

　　在經過一段車程後，你抵達公司上班了。在到晚間六點下班前的這段時間，我想你應該可以全心為工作規劃時間。但是，在中午用餐的這一個或一個半小時裡，午餐時間實際上並不用到半小時，總之，這是完完全全屬於自己的時間，你可以充分地支配使用，不論是休息、和同事聯絡感情、培養興趣、整理周圍的環境……甚至讀報也是可以的。

好不容易等到下班了！這時候的你，如果依照孩子們的說法：你看起來簡直累得像頭牛！不但精神不濟、滿臉倦容，恍惚的神情就好像是慘遭過疲勞轟炸的憔悴。在回家的路上，你的慢性疲勞症越發地加重了。身心上的疲憊不堪，就像黑鴉鴉的烏雲，沉甸甸地壓垮了你快要無法負載的身軀，又彷彿像是深冬寒夜裡的冷風，凍得讓你都沒了知覺。

好不容易回到家，雖然有親人精心準備的美味佳餚，你卻食不知味，滿腦子只想囫圇吞棗完後放空自己。沉默無語地休息一個小時後，你的精神開始恢復了些許，這時你或許會想吃點甜品、抽根菸、和家人聊聊天、打打小牌，或者出外遛遛狗、逛逛街，再玩玩樂器。當你抬頭看看時鐘，喔！已經十一點十五分了（晚間）。

由於擔心一上床又要失眠的問題，於是你又想著要不要晚點睡覺呢？就這樣猶豫不決間又過了四十分鐘。你決定喝點威士忌再上床，但是卻始終輾轉難眠。終於你開始有些睡意了，這一天也可以畫上句號了。回想一下吧！自你下班後到上床睡覺之前的這六個小時（或許更久），是不是像做了場夢般地，倏忽就消失呢？

以上的描繪，是否讓許多人也有著某種熟悉感？事實上，它就是很多人下了班的故事。或許有人會想反駁：「你說的可能都是真相，但沒辦法啊！工作一天後，我就是很累嘛！但又不能讓壓力綁死我吧！因為想要放鬆，所以才和朋友打打麻將，喝喝小酒，娛樂一下也是生活中的必要啊！」

乍聽之下，似乎頗為合理。但我們若假設另一種狀況，那又會如何變化呢？好，我們預定你和情人約在晚上八點去看電影，你決定趁下班到約會前的空檔，回家梳洗換裝一番。於是你火速地搭上計程車，完成任務後，又飛快地搭上另一輛計程車，奔往電影院和情人會合。

在你們相處的整個晚上，至少有不少於四個小時的時間吧！你的心情始終是亢奮而神采奕奕的。終於，到了依依不捨分手回家的時刻。一回家後，你簡潔快速地處理完所有私事，就準備上床做一夜好夢。

這一晚，你竟然可以不用像平日一樣，花上近一個小時磨磨蹭蹭地才肯睡覺，想必是因為甜蜜蜜的快樂情緒，趕走你工作一整天的疲勞和苦悶壓力吧。這一個晚上，即使是長達將近六個小時的奔波忙碌，你還是意猶未盡地回味著美好時光！

這是否勾起你另一個回憶，那年你接受了朋友的建議，加入一個業餘合唱團的演出。在三個月的訓練期間，每隔一個晚上你就要用至少兩個小時的時間參加培訓，雖然過程很辛苦，但卻樂此不疲。因為那段期間讓你覺得，在無趣的生活中增添了許多色彩，似乎又重燃起你對生命的熱情。

由以上的兩個例子，我想向讀者證明的是，你必須真誠地面對自己。事實上，在六點下班後的這段時間，你的狀況並沒有如想像中的疲憊不堪。如此一來，你才可以完整地規劃下班後的活動，而不會因為疲倦導致被未能達到效果的休息所中斷。若是以吃完晚餐到睡覺前的時間計算，至少有不少於三個小時的時間，可以讓你充分地補充能量。

　　我並非要你下了班後還必須繼續讓腦不斷地「工作」，而是希望你能至少每隔一個晚上，用一個半小時來動動腦想想「工作」以外的事，活化刺激一下將被壓力淹沒的創意，或者思考人生的方向及生活周遭的一切……等。

　　亦即，**撥出一段重要的時間來思考生命中重要的事**。之後，你仍然有兩個晚上及整整兩個假日的時間，至少有將近四十個小時以上的財富，可以用來從事其他的休閒活動，諸如和朋友敘舊、打球、學烹飪、散步、開車兜風……等。

　　萬事起頭難，但只要你將眼光放遠，這是為了成就人生所必要的準備。只要堅持到了第四個，甚至第五個晚上，你忙亂的身心將逐漸安頓。這時你將發現，原來，自己也可以有這樣的生活方式。就算不知不覺中又到了該睡覺的時間，你也不會再輾轉反側了。

　　事實上我們平日最浪費的時間，就是花在想睡又拖拖拉拉的那段零碎時間。反反覆覆地閉著眼，張開眼，腦裡偶爾一片空白，偶爾又亂七八糟地想著事，這樣的時間加加總總，都可能可以看完一本中篇小說了。重要的是，生命就在無意義中又消逝了一個小時。

請謹記在心，最初就要將這每隔一天的一個半小時，視同如一星期中每一分每一秒一樣的重要。保持著你和情人約會或參加合唱比賽的心情，興致勃勃、努力不懈地持續下去。

千萬別再像以前一樣，總在中途就放棄：「抱歉，我今天可能要缺席一次，因為朋友等我去打球呢！」此刻，請你要認真地對自己說：「不行的，如果要讓生活有意義，一次學習也不能停止。」或許有些困難，但若讓一時的玩樂不斷地上了癮，你的人生，大概再也沒有快樂起來的理由吧！

Chapter V

Tennis and the Immortal Soul

You get into the morning train with your newspaper, and you calmly and majestically give yourself up to your newspaper. You do not hurry. You know you have at least half an hour of security in front of you. As your glance lingers idly at the advertisements of shipping and of songs on the outer pages, your air is the air of a leisured man, wealthy in time, of a man from some planet where there are a hundred and twenty-four hours a day instead of twenty-four. I am an impassioned reader of newspapers. I read five English and two French dailies, and the news-agents alone know how many weeklies, regularly. I am obliged to mention this personal fact lest I should be accused of a prejudice against newspapers when I say that I object to the reading of newspapers in the morning train. Newspapers are produced with rapidity, to be read with rapidity. There is no place in my daily programme for newspapers. I read them as I may in odd moments. But I do read them. The idea of devoting to them thirty or forty consecutive minutes of wonderful solitude (for nowhere can one more perfectly immerse one's self in one's self than in a compartment full of silent, withdrawn, smoking males) is to me repugnant. I cannot possibly allow you to scatter priceless pearls of time with such Oriental lavishness. You are not the Shah of time. Let me respectfully remind you that you have no more time than I have. No newspaper reading in trains! I have already "put by" about three-quarters of an hour for use.

Now you reach your office. And I abandon you there till six o'clock. I am aware that you have nominally an hour (often in reality an hour and a half) in the midst of the day, less than half of which time is given to eating. But I will leave you all that to spend as you choose. You may read your newspapers then.

I meet you again as you emerge from your office. You are pale and tired. At any rate, your wife says you are pale, and you give her to understand that you are tired. During the journey home you have been gradually working up the tired feeling. The tired feeling hangs heavy over the mighty suburbs of London like a virtuous and melancholy cloud, particularly in winter. You don't eat immediately on your arrival home. But in about an hour or so you feel as if you could sit up and take a little nourishment. And you do. Then you smoke, seriously; you see friends; you potter; you play cards; you flirt with a book; you note that old age is creeping on; you take a stroll; you caress the piano.... By Jove! a quarter past eleven. You then devote quite forty minutes to thinking about going to bed; and it is conceivable that you are acquainted with a genuinely good whisky. At last you go to bed, exhausted by the day's work. Six hours, probably more, have gone since you left the office--gone like a dream, gone like magic, unaccountably gone!

That is a fair sample case. But you say: "It's all very well for you to talk. A man *is* tired. A man must see his friends. He can't always be on the stretch." Just so. But when you arrange to go to the theatre (especially with a pretty woman) what happens? You rush to the suburbs; you spare no toil to make yourself glorious in fine raiment; you rush back to town in another train; you keep yourself on the stretch for four hours, if not five; you take her home; you take yourself home. You don't spend three-quarters of an hour in "thinking about" going to bed. You go. Friends and fatigue have equally been forgotten, and the evening has seemed so exquisitely long (or perhaps too short)! And do you remember that time when you were persuaded to sing in the chorus of the amateur operatic society, and slaved two hours every other night for three months? Can you deny that when you have something definite to look forward to at eventide, something that is to employ all your energy--the thought of that something gives a glow and a more intense vitality to the whole day?

What I suggest is that at six o'clock you look facts in the face and admit that you are not tired (because you are not, you know), and that you arrange your evening so that it is not cut in the middle by a meal. By so doing you will have a clear expanse of at least three hours. I do not suggest that you should employ three hours every night of your life in using up your mental energy. But I do suggest that you might, for a commencement, employ an hour and a half every other evening in some important and consecutive cultivation of the mind. You will still be left with three evenings for friends, bridge, tennis, domestic scenes, odd reading, pipes, gardening, pottering, and prize competitions. You will still have the terrific wealth of forty-five hours between 2 p.m. Saturday and 10 a.m. Monday. If you persevere you will soon want to pass four evenings, and perhaps five, in some sustained endeavour to be genuinely alive. And you will fall out of that habit of muttering to yourself at 11.15 p.m., "Time to be thinking about going to bed." The man who begins to go to bed forty minutes before he opens his bedroom door is bored; that is to say, he is not living.

But remember, at the start, those ninety nocturnal minutes thrice a week must be the most important minutes in the ten thousand and eighty. They must be sacred, quite as sacred as a dramatic rehearsal or a tennis match. Instead of saying, "Sorry I can't see you, old chap, but I have to run off to the tennis club," you must say, "...but I have to work." This, I admit, is intensely difficult to say. Tennis is so much more urgent than the immortal soul.

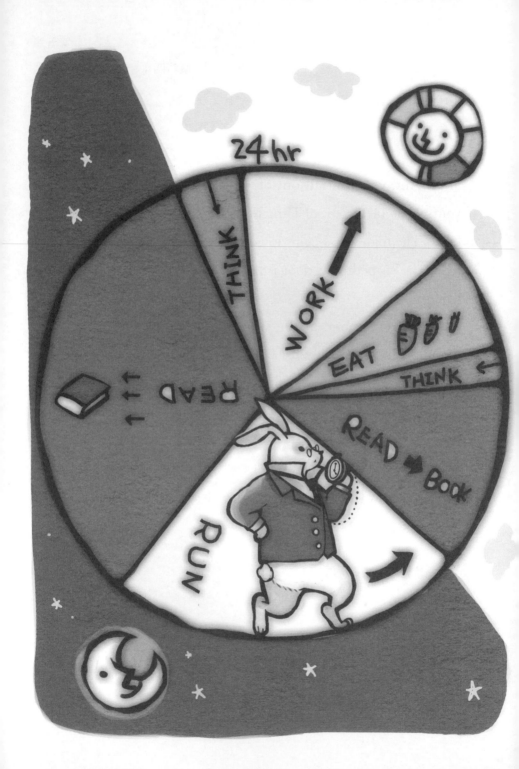

第六章

策略五：合乎人性的時間管理

不要以為假日或下班後的時間，是我們必然的所得。
這些「多餘的時間」很容易讓人漫不經心地就揮霍了。
將它視為從天而降的禮物，為自己的人生做有意義的
計畫及使用。

　　現代人真的很幸福，由於各種法制的保障及福利的爭取，比起以前的勞動者，每天動輒要做滿十個小時的工作，現在還多了兩個假日的休息，整整四十八個小時可供我們完全利用。但是實際上，有許多同時被工作與生活趕得焦頭爛額的上班族來說，他們長久以來還是一直有個疑惑，他們的一週究竟是五天還是七天呢？

　　事實上包括我自己，即使到了四十歲時，我一週的工作時間的確還是七個整天。雖然有許多前輩經常殷殷勸說，要我力行一週至少休息一天，如此不但會使工作的品質更加提升，也能讓自己從壓力中解脫而煥然一新，才能精神加倍地面對下週的任務。

　　或許是因為年紀漸長吧，我才能開始深刻地體會到前輩們說的好處。現在的我，已經不會再讓自己在整個七天都無時無刻地沉沒在工作中，也不會再神經緊繃地預備職場的突發狀況。一週中至少有一天，我會給自己完整而放鬆的休息。

　　可是說實話，若是人生可以重來一次，我還是會選擇努力不懈地工作七天。因為我認為，只有那些長久以

來每天都用心於工作的人們，他們才能瞭解休假的珍貴
用意，也才有資格獲得假日。

更何況，對於我們這些逐漸邁入銀髮族的人來說，
當看見那些活力充沛的年輕人奮不顧身地工作時，更要
喟嘆果真歲月如梭！因此眼見流逝的青春不再，我就更
忍不住心酸地想對那些花樣年華的小夥子們說：「衝啊！
為你的人生勇往直前吧！」至少，我個人是如此深切地
渴望著。

雖然最好是盡量在上班的五天中，將一週的工作完
成，但是，若是遇到有些工作狂或是樂在工作者，我也
只能俯首稱臣地說：那就盡其所能地利用這屬於你的假
日時光，工作吧！只是請務必要量力而為，如果覺得身
心難以負荷，就要懂得放手或授權。要知道，學會適度
地放鬆自己，也是生命中很重要的課程。

此外，不要以為假日或下班後的時間是我們理所當
然的獲得。因為這些「多餘的時間」，很容易就會讓人
在漫不經心中揮霍了。要將它視為從天而降的禮物，為
自己的人生做有意義的計畫及使用。如此一來在接續而

來的一週裡，你會因為完整的充電或休息，而更加倍精神地全力以赴於工作及生活上。

好吧！可能有人會質疑：如果扣除掉週休二日及下班後的時間，加上生活中還要做一堆拉拉雜雜的私人瑣事，我們究竟還有多少「多餘的時間」，是可用來作為我所謂的「人生投資」呢？

讓我來加總一下那些不該浪費的時間吧：早上等車和坐車的零碎時間約有三十分鐘，再加上上一章我所建議的，每隔一日取用的一個半小時，合計共為七個小時。看吧！你在一星期裡最少有七個小時的黃金時間，可以用來投資你的人生。

或許又有人要反彈了：「不會吧，你不是要教我們掌控一天二十四小時，現在卻抽絲剝繭似的在一週一百六十八小時裡，算出零零總總的七個小時，什麼嘛！七個小時就可以營造出生命的奇蹟嗎？怎麼可能呢！」

沒錯，親愛的朋友們，別懷疑，事實就是如此。

千萬別小看這每週七個小時的成果！重要的是你

務必要來親身體驗一下，這七個小時所能創造的生命奇
蹟。也就是說，如果你能善加利用這一週的七個小時，
你不但可以以最高效率完成一週的任務，你再也不用在
週日晚間就受「工作倦怠症」所苦，你的人生從此以後
也會充滿更多的樂趣。這樣的生命奇蹟，難道不是我們
日夜所渴求的嗎？

　　於是又有人問：「那麼，這七個小時，我又該如何
利用呢？」這是很簡單的。只要你願意開始實行，即使
只是每天早起做十分鐘的運動，長久下來後，你會發現
自己的體質正在改變，而所有的情況也都在逐漸好轉。

　　或者，你正受困於生活上的種種煩惱，那何不每天
撥出一個小時，跟自己的心靈對談呢？慢慢地，原本陰
暗枯萎的心園，又會感覺到綠意盎然的生機，問題自然
迎刃而解。你要相信自己，永遠是自己最好的心理輔導
師。

　　當然有人會認為，應當將更多的時間用在投資自己
的身上，這也絕對是一件穩賺不賠的投資。更何況對自
己投資得越多，相對的，未來人生的成果應當更加豐富。

　　但是要記得築夢踏實，若能利用這每週七個小時將根基奠定，未來的發展才能穩紮穩打。

　　從另一個角度來看，要能從一週中撥出七個小時自我投資，其實並非是件容易的事。因為那代表著，你必須要付出某些代價，亦即某種程度的犧牲。一開始，你或許將時間利用得支離破碎，始終沒能有一件成果出現；也或許在那段時間，你感到窮極無聊而不知所措。但沒關係，這些都只是過程，因為你正在學習著改變習慣，必然會產生某些不適應的狀況，但終究還是可以渡過難關。

　　你我都知道，要改變習慣可不是件輕鬆的事，更別說是在改變的歷程中，經常會出現許多矛盾、挫折及衝擊。正如自然產的新生兒，總免不了讓母親經歷永生難忘的劇痛。

　　所以要是有人以為只要每週撥出七個小時，就能換得未來的精彩人生，那可就大錯特錯了！別忘了，我一再強調的：你要有所付出，你要有所犧牲，你更要努力堅持，要記得為了讓自己脫胎換骨，重新擁有想要的人生，這一切都是值得的。

　　但是，事情偶爾可能並未如想像中的容易辦到，而且一不留心，很可能會將自己陷於失敗的萬劫不復之中。所以我們可以從最簡單的地方，或認為最有趣的方法開始，例如每天五分鐘的甩手功、十分鐘的發音練習、十五分鐘的穴道按摩、三十分鐘的單字背誦……等。總之，要時時保持警覺，別因自滿而產生鬆懈，以至於遭受挫敗而潰不成軍。

　　除此之外，要想讓有益自己的事能持之以恆地進行，自信是非常重要的。即使規劃再好的計畫，只要中途遇到一兩次的挫折，都會讓人感到飽受打擊的痛苦，而輕易地宣告放棄。所謂「欲速則不達」，操之過急往往最容易造成功虧一簣的狀況。

　　最後，只要給自己三個月的時間──若是你能確確切切地落實，每週撥出七個小時自我投資，你就可以清清楚楚地感受到自己的改變，並發現生活周遭有了神奇的變化。

　　我要不厭其煩地提醒讀者！在進行練習前，要記得多預留一些時間以防意外發生。由於一開始，我們尚未能掌握運用自己時間的訣竅，因此難免會受到干擾而有所影響，所以在安排晚間的一個半小時的時候，可將時間再略為延長，如原本預定為八點到九點半，則可改到十點半結束，才能學習如何完整地使用這一個半小時。

Chapter VI
Remember Human Nature

I have incidentally mentioned the vast expanse of forty-four hours between leaving business at 2 p.m. on Saturday and returning to business at 10 a.m. on Monday. And here I must touch on the point whether the week should consist of six days or of seven. For many years--in fact, until I was approaching forty--my own week consisted of seven days. I was constantly being informed by older and wiser people that more work, more genuine living, could be got out of six days than out of seven.

And it is certainly true that now, with one day in seven in which I follow no programme and make no effort save what the caprice of the moment dictates, I appreciate intensely the moral value of a weekly rest. Nevertheless, had

I my life to arrange over again, I would do again as I have done. Only those who have lived at the full stretch seven days a week for a long time can appreciate the full beauty of a regular recurring idleness. Moreover, I am ageing. And it is a question of age. In cases of abounding youth and exceptional energy and desire for effort I should say unhesitatingly: Keep going, day in, day out.

But in the average case I should say: Confine your formal programme (super-programme, I mean) to six days a week. If you find yourself wishing to extend it, extend it, but only in proportion to your wish; and count the time extra as a windfall, not as regular income, so that you can return to a six-day programme without the sensation of being poorer, of being a backslider.

Let us now see where we stand. So far we have marked for saving out of the waste of days, half an hour at least on six mornings a week, and one hour and a half on three evenings a week. Total, seven hours and a half a week.

I propose to be content with that seven hours and a half for the present. "What?" you cry. "You pretend to show us how to live, and you only deal with seven hours and a half out of a hundred and sixty-eight! Are you going to perform a miracle with your seven hours and a half?" Well, not to mince the matter, I am--if you will kindly let me! That is to say, I am going to ask you to attempt an experience which, while perfectly natural and explicable, has all the air of a miracle. My contention is that the full use of those seven-and-a-half hours will quicken the whole life of the week, add zest to it, and increase the interest which you feel in even the most banal occupations. You practise physical exercises for a mere ten minutes morning and evening, and yet you are not astonished when your physical health and strength are beneficially affected every hour of the day, and your whole physical outlook changed. Why should you be astonished that an average of over an hour a day given to the mind should permanently and completely enliven the whole activity of the mind?

More time might assuredly be given to the cultivation of one's self. And in proportion as the time was longer the results would be greater. But I prefer to begin with what looks like a trifling effort.

It is not really a trifling effort, as those will discover who have yet to essay it. To "clear" even seven hours and a half from the jungle is passably difficult. For some sacrifice has to be made. One may have spent one's time badly, but one did spend it; one did do something with it, however ill-advised that something may have been. To do something else means a change of habits.

And habits are the very dickens to change! Further, any change, even a change for the better, is always accompanied by drawbacks and discomforts. If you imagine that you will be able to devote seven hours and a half a week to serious, continuous effort, and still live your old life, you are mistaken. I repeat that some sacrifice, and an immense deal of volition, will be necessary. And it is because I know the difficulty, it is because I know the almost disastrous effect of failure in such an enterprise, that I earnestly advise a very humble beginning. You must safeguard your self-respect. Self-respect is at the root of all purposefulness, and a failure in an enterprise deliberately planned deals a desperate wound at one's self-respect. Hence I iterate and reiterate: Start quietly, unostentatiously.

When you have conscientiously given seven hours and a half a week to the cultivation of your vitality for three months--then you may begin to sing louder and tell yourself what wondrous things you are capable of doing.

Before coming to the method of using the indicated hours, I have one final suggestion to make. That is, as regards the evenings, to allow much more than an hour and a half in which to do the work of an hour and a half. Remember the chance of accidents. Remember human nature. And give yourself, say, from 9 to 11.30 for your task of ninety minutes.

第七章
策略六：注意力是成就的關鍵

每天的第一項工作，就是先練習提高集中專注力，
意識到自己要開始真正的生活。
就像有人一早起來，總先注意自己的外表儀容；
有些人為了美觀會不惜忍痛除毛，這都是一樣的道理。

有人說：「人類不能掌控自己的思想。」事實果真如此嗎？我們的大腦不僅可以產生各種的想法，快樂的、痛苦的、憤怒的各種情緒，也都是經由在腦海裡翻轉而表現於外，所以我們當然可以左右自己的思想。

但上述這句話真正的涵義，並非如表面文字所呈現的意思。實際上，它是要我們反躬自省：在我們之中有多少人曾真正地、完全地掌握過自己的思想呢？現實生活中，我們若真能做到不受到負面思想的影響，就已經能讓自己受用一生了。

相對的，大部分的人終其一生總是怨天尤人。我們總是輕易地將責任推卸到別人身上，老是指著別人的不是說：都是你啦！才會導致事情演變不利……等等。所謂：「當你一指指向別人，別忘了，還有四指指向自己。」我們應該意識到，這一切都是我們的決定，若是我們當初選擇相信自己的判斷，別人又何必要成為你脫罪的藉口呢？

想要讓自己擁有正確的判斷力，就要先學會集中注意力。因為，你若無法使大腦聽命行事，它就會如脫韁野馬地不受控制，這麼一來，你又要如何看清事情的本質及問題的所在呢？所以，你要先學會掌控你的意志，才能感受到生活真實的存在。

以我來說，每天的第一項工作，就是要先練習提高集中專注力，意識到自己要開始真正的生活。就像有人一大清早起來，總會先注意自己的外表儀容；有些人為了美觀，不惜忍住疼痛而除毛；為了獲得更多的健康，有人會和小販、直銷商、營養師、醫師不斷地拉關係，攀人情，但卻少有人想到或多關心一下，自己的思想——這個整天在為我們拚命下決定的「機器」。

因此，我每天都會將這段時間儲蓄起來——從出了家門口的那一刻，到抵達公司的那一刻，那是一段培養意志的練習之旅。在不需要任何人的協助下，我一個人就能學習的人生技巧。

「什麼？要我在大街上、車站邊、電車裡，在人來人往中學會集中注意力嗎？」不要懷疑，雖然不可能輕鬆地就學會這種技巧，但事情絕對非你想像中的困難。畢竟，你完全不需要借助任何一項工具，更無須為此去研讀任何一本書。

　　只要當你一離開家後，你就開始將注意力集中去想一件事，任何一件事都可以。大約走不到五分鐘後，你會發現注意力已經逐漸在動搖之中，漸漸地，出現在你身旁的事物都可以輕而易舉地誘引它。這時，你要像拉住不聽使喚的小狗，將牠引導回本來的位置。

　　在你抵達車站前，單是這樣的動作，可能反反覆覆地就要來上數十遍。千萬不用氣餒，堅持不懈終究會使你邁向成功。相反的，你若在此時就放棄了，說什麼你就是無法集中注意力；說什麼花花世界，很難讓人不受影響，說什麼……別再說太多藉口了，我想你還是承認吧，歸根究柢只有一個原因：你太懶散了！

　　不妨，在記憶箱裡找尋一下經驗吧！某天早上，當你打開電子信箱收信，卻意外地收到一封來自廠商的壞消息。你感到渾身不自在，此刻，你只想到要如何謹慎地回信。於是從出家門到抵達辦公室的這一段時間，你無時無刻不在思量著回信的內容。終於你快速地坐在辦公桌前，打開電腦，完全無視他人的存在，拚命地、飛快地打著回信。

　　現在，讓我們來檢視一下你在這件事的表現：關於你對自己的注意力集中，簡直優秀到可圈可點的程度。你排除所有的雜念，就像一個專制的君主，獨裁地控制著自己的思想。於是你辦到了，你完完全全地將注意力掌控在自己手中，只因為有志者事竟成。因此除了持之以恆外，要完成注意力集中的練習並沒有其他的技巧。透過日復一日、堅持不懈的訓練，完全不須攜帶任何器具便可以簡單地達成目的。

　　想想看，你若是為了要鍛鍊肌肉，每天早晨還要拎著啞鈴去公園；為了要參加益智比賽，也要猛Ｋ百科全書。唯獨訓練你的意志力，不管是靜悄悄地站在公車站等車，還是抓著吊環在公車上搖來晃去，完全沒有人能察覺到，你正在默默地學習注意力集中，更別擔心中途失敗了會招來恥笑。

　　再一次地提醒，我在意的是你要學會集中注意力，而不是為了集中注意力所想的是什麼。這是一種對自我思想的控制練習，並不著重於所想的問題。就像你可以

聚精會神地想著馬可‧奧理略（Marcus Aurelius，註 1）
或是伊比鳩魯（Epicurus，註 2）的某節作品。當然，如
果你所想的事是正念而有意義的，那你所得到的好處就
更多了。

　　無須擔心自己要如何看懂馬可‧奧理略或是伊比鳩
魯的名著，我只是借他們兩位來比喻鼓舞人心的文章；
亦即，你可以多看些激勵人性的作品，即使是五六百字
的極短篇也可以。在每天晚上及次日的清晨，多品讀幾
次，等時間久了以後，你就會體認到我所說的深奧之處。

　　親愛的朋友，看到此處或許有人又想舉手反對：
「好吧！我承認，到第七章，我都可以認同你的意見，
但是，你要我在等車時練習注意力集中，我覺得有點不
可思議，礙難照辦！」

　　先別急著拒絕這項建議吧！或許它看起來有些高難
度，也或許是有些可笑，但這可不是我先想到的方法，
早在許久前，就有許多先知先覺一而再，再而三地提醒
過我們：「要掌握自己的思想，才能掌握自己的人生。」
我只不過是他們的轉述者。

　　無論如何，請你排除偏見，努力嘗試看看吧！當你在學習集中注意力的同時，你會發現另一個意外的收穫，就是那個老是讓你遭受突發性災難的罪魁禍首——恍神小姐，她從此也將跟你的人生告別！

──────────

註 1 馬可・奧理略（Marcus Aurelius，121 － 181 年），被稱為「哲學家皇帝」，是羅馬帝國最偉大的皇帝之一，著有《沉思錄》一書。

註 2 伊比鳩魯（Epicurus，前 341 －前 270 年），古希臘哲學家，相信無神論，並認為善來自於快樂。《伊比鳩魯悖論》是其著名遺產之一。

Chapter VII
Controlling the Mind

People say: "One can't help one's thoughts." But one can. The control of the thinking machine is perfectly possible. And since nothing whatever happens to us outside our own brain; since nothing hurts us or gives us pleasure except within the brain, the supreme importance of being able to control what goes on in that mysterious brain is patent. This idea is one of the oldest platitudes, but it is a platitude who's profound truth and urgency most people live and die without realising. People complain of the lack of power to concentrate, not witting that they may acquire the power, if they choose.

And without the power to concentrate--that is to say, without the power to dictate to the brain its task and to ensure obedience--true life is impossible. Mind control is the first element of a full existence.

Hence, it seems to me, the first business of the day should be to put the mind through its paces. You look after your body, inside and out; you run grave danger in hacking hairs off your skin; you employ a whole army of individuals, from the milkman to the pig-killer, to enable you to bribe your stomach into decent behaviour. Why not devote a little attention to the far more delicate machinery of the mind, especially as you will require no extraneous aid? It is for this portion of the art and craft of living that I have reserved the time from the moment of quitting your door to the moment of arriving at your office.

"What? I am to cultivate my mind in the street, on the platform, in the train, and in the crowded street again?" Precisely. Nothing simpler! No tools required! Not even a book. Nevertheless, the affair is not easy.

When you leave your house, concentrate your mind on

a subject (no matter what, to begin with). You will not have gone ten yards before your mind has skipped away under your very eyes and is larking round the corner with another subject.

Bring it back by the scruff of the neck. Ere you have reached the station you will have brought it back about forty times. Do not despair. Continue. Keep it up. You will succeed. You cannot by any chance fail if you persevere. It is idle to pretend that your mind is incapable of concentration.

Do you not remember that morning when you received a disquieting letter which demanded a very carefully-worded answer? How you kept your mind steadily on the subject of the answer, without a second's intermission, until you reached your office; whereupon you instantly sat down and wrote the answer? That was a case in which *you* were roused by circumstances to such a degree of vitality that you were able to dominate your mind like a tyrant. You would have no trifling. You insisted that its work should be done, and its work was done.

By the regular practice of concentration (as to which there is no secret-save the secret of perseverance) you can tyrannise over your mind (which is not the highest part of *you*) every hour of the day, and in no matter what place. The exercise is a very convenient one. If you got into your morning train with a pair of dumb-bells for your muscles or an encyclopaedia in ten volumes for your learning, you would probably excite remark. But as you walk in the street, or sit in the corner of the compartment behind a pipe, or "strap-hang" on the Subterranean, who is to know that you are engaged in the most important of daily acts? What asinine boor can laugh at you?

I do not care what you concentrate on, so long as you concentrate. It is the mere disciplining of the thinking machine that counts. But still, you may as well kill two birds with one stone, and concentrate on something useful. I suggest--it is only a suggestion--a little chapter of Marcus

Aurelius or Epictetus.

Do not, I beg, shy at their names. For myself, I know nothing more "actual," more bursting with plain common-sense, applicable to the daily life of plain persons like you and me (who hate airs, pose, and nonsense) than Marcus Aurelius or Epictetus. Read a chapter--and so short they are, the chapters! --in the evening and concentrate on it the next morning. You will see.

Yes, my friend, it is useless for you to try to disguise the fact. I can hear your brain like a telephone at my ear. You are saying to yourself: "This fellow was doing pretty well up to his seventh chapter. He had begun to interest me faintly. But what he says about thinking in trains, and concentration, and so on, is not for me. It may be well enough for some folks, but it isn't in my line."

It is for you, I passionately repeat; it is for you. Indeed, you are the very man I am aiming at.

Throw away the suggestion, and you throw away the

most precious suggestion that was ever offered to you. It is
not my suggestion. It is the suggestion of the most sensible,
practical, hard-headed men who have walked the earth.
I only give it you at second-hand. Try it. Get your mind
in hand. And see how the process cures half the evils of
life -especially worry, that miserable, avoidable, shameful
disease-worry!

第八章
策略七：信念影響命運

攸關人生的問題，我們都未曾深刻地思考過。
關於我們的幸福、我們的未來、生活中的擁有及失
去……等，
許許多多與人生密不可分的事情，
我們總是睜一隻眼，閉一隻眼地盡可能逃避閃躲！

今天，你做練習了嗎？就如正式學鋼琴前的音階基本課程，每一天，你都務必要花三十分鐘以上，學習如何使注意力高度集中。

我們必須先學會掌控這人體中最複雜、最難以駕馭的器官——大腦，才能進一步供給它有意義的養分（思想）。相對的，如果光只能對大腦控制自如，卻沒有使它獲得有益的想法，那就成了空轉的機器，一切都是徒勞無功。因此接下來，我們將進階地來研究，何謂有意義的思想。

對於我們的人生來說，這是一門漫漫長路卻又無比重要的課程。它並非是要探討文學或是藝術，也不是關於任何一種科學。事實上，它就是和我們自己息息相關——一門關於人類的研究。

有一句古老的諺語：「人類啊！你要懂得瞭解自己。」或許聽起來是老生常談，也或許有人認為有點陳腔爛調，但是請容我向各位再一次地強調，「瞭解自己」真的很重要。

綜觀人類數萬年的文明發展，科技的進步日益月新，但唯獨對自己的認識卻還是迷迷糊糊的，得過且過。

數千萬人中，也僅有那少得不能再少的聖賢先知能問心無愧地說：「人生啊！我當盡力而為。」而大部分的人，當然也包括你我，捫心自問一下吧，我們何時曾認認真真地思考過「何謂人生」，抑或是，我們要度過怎樣的人生呢？

這些攸關人生的問題，我們都未曾深刻地思考過。關於我們的幸福、我們的未來、生活中的擁有及失去、為何做了這樣的事、為何有那樣的想法……等，許許多多與人生密不可分的事情，我們總是睜一隻眼，閉一隻眼地盡可能逃避閃躲，或是毫無知覺地面對著它們。

生活，並非只是活著就好。你要在生活中感受生命存在的價值，也要在生命中體驗有意義的生活。這不就是每個人都想尋尋覓覓所謂的「幸福」嗎？那麼，你有找到嗎？或許你還在尋找，也或許你認為幸福只是遙不可及的謊言，所以放棄了。但是別忘了，在這世上，依然有人找到自己的幸福了。

是的，他們真的得到了幸福。那是因為他們瞭解幸福的真諦，並非是物質上或精神上的一時快樂。它是以智慧面對人生，願意相信生命中的一切，並全心地為它付出，它使我們的生命獲得永恆的喜樂。

我想任何人都很難否認以上的事實吧！既然如此，何不在每一天多花些時間，來檢討自己的想法、行為及信念呢？想想我們的做事態度吧！即使是認真地全力以赴，但為何不是丟三落四，就是橫生枝節？這一切究竟是別人造成的禍害，還是自己哪裡出了問題？現在，你知道是誰才該好好思考了吧！

你無須排斥，請別以為我要灌輸新的信念。當下，你的想法是如何，其實並不重要。就算你滿腦子想入非非，負面的、小器量的種種念頭，我都不會感到訝異，真正重要的是你的想法和你的行為是否一致。

這個社會有太多言行不一的人，說的和想的不一樣，想的和做的又不同，甚至連說出來的話都不一定會做。這樣的人，每天、每時、每刻都在衝突和矛盾中掙扎，反覆無常的信念，讓他們生活得很辛苦，很不快樂。事實上，他們經常會質疑自己的人生是否真有意義呢？

請先撇開道德的看法，我想再進一步向各位證明，信念和行為的關聯性。假若Ａ君認為助人為快樂之本，即使劫富濟貧也是可以犯的錯，因此就算到最後，他因為犯罪而必須身陷囹圄，但只要一想到曾受他幫助的人已脫離苦海，即便如此他依然是甘之如飴。

再反觀Ｂ君，他可說是個極為潔身自愛的人，但卻因貧病交迫而「不得不」觸犯法律，最後他當然被捕入獄，但是從此他也將自己判了終身監禁。事實上，真正懲罰到他的，不是法官判了有期徒刑，也不是指指點點的輿論，而是他的良知，是他賴以生存的信念──凡人都不應該殺搶偷騙，否則即為神的罪人。

因此，殉道者的道理也是一樣的。在別人眼裡看來，他們可能是一群不可思議的盲目傻瓜，但他們卻不以為然。只要是為了實踐至高的信念，即使犧牲了生命，但他們卻到達生命中夢寐以求的極樂天堂，這是他們人生中的夙願，他們當然引以為傲。

大家都公認，人類絕對是這世上最有理性的生物，但實際上，我們在大多數時候的行為卻是受「感覺」的

影響，而非理性的判斷。因此雖然有人說：「理智是行
為的依據。」但在真實的生活中，它卻經常沒能發揮太
大效果，更別說理智對如影隨形的信念有任何作用。

　　但無庸置疑的，越不曾經過反省的理智，其功能就
會越來越降低。舉例來說，你某天興高采烈地到餐廳想
要大啖美食一番，不料卻被服務生端來的過熟牛排完全
壞了整天的興致。當你怒不可遏，正要對服務生破口大
罵時──可否請先慢個兩秒鐘，你的理智，想要告訴你
一件事實：

　　「喔，請先別動怒。想想看吧，那服務生也真夠倒
楣，那份牛排又不是他搞砸的，卻要莫名其妙地遭人一
頓罵。好吧，就算你真罵了他，那也無濟於事啊！牛排
既不會變好吃，你的心情更不會因此變好，更何況別的
客人會怎麼看你？或許他們會說：『瞧瞧那傢伙，有必
要這樣欺負那個可憐的小服務生嗎？』『喔！可真是得
理不饒人。』」

你認真地思量一下，還真是很不划算吧？所幸，最後你還是採用了理智的建議。想想看，這傢伙（理智）還真是好用，它既不會向你收諮詢費，也不會因此嘲笑你的無知魯莽，接下來，它還會引導你未來要如何處理：

「萬一，下一次你又遇到相同的狀況，請先別急著對那個可憐的小服務生生氣。你可以深呼吸，讓自己保持平靜，然後，溫和有禮地請服務生更換一盤牛排。如此一來，你非但不會破壞了自己美好的用餐心情，也不會使其他人受到無辜的牽連，是不是一舉數得呢？」

想必從此以後，你一定更會善用理性的好處，也會因此減少許多處理不必要發生的麻煩的時間了。

另一方面，對於想要建立正確信念並決定將其兌現的人，可以試著選擇閱讀一些世界名著，或賢人名家所寫的勵志小品。如上一章曾提及的馬可・奧理略或伊比鳩魯。當然，我也樂於推薦如帕斯卡（Pascal，註1）、拉布魯耶（La Bruyere，註2）或是愛默生（Emerson，註 3）的作品也相當適合。

書，對我而言誠然是無價之寶。即使在旅程中，我也總會帶著如馬可‧奧理略等人的著作隨時翻閱。但是這並非要你用閱讀，取代所有的自我訓練。你仍然要每天誠實地面對，並反省自己的一切言行。即使，檢視的眼神讓你會有所愧疚，你還是絲毫不能鬆懈。

至於反省自我的工作，要何時進行最為恰當呢？我的建議，下班後回家的路上是個不錯的時機。因為在辛苦工作了一天後，如果可以靜心回首今天一日的功過，然後給自己一個愛的訓話及激勵，之後更能自在無虞地放鬆及專心地自我訓練。

或許有人認為，下班後他只想做些輕鬆的活動，讓自己整天動個不停的腦子好好休息。那麼，不管是想讀報或是任何其他的方式，都請隨心所欲地採用。最要緊的是，由於反省自我是檢視生命的重要功課，所以請你每一天，不管在哪一個時刻都要抽空完成此項課程，長久下來，你一定可以看見生命的巨大改變。

––––––––––

註 1 布萊茲‧帕斯卡（Blaise Pascal，1623 － 1662 年），法國數學家、物理學家及思想家，早期主要研究自然和應用科學，對機械計算器的製造和液體的研究做出重要貢獻，並提出著名的「帕斯卡定律」。之後轉為專注於沉思和神學、哲學的寫作。

註 2 拉布魯耶（La Bruyere，1645 － 1696 年），法國哲學家和作家，以深刻洞察人生的著作《品格論》而聞名。

註 3 愛默生（Ralph Waldo Emerson，1803 － 1882 年），是十九世紀美國重要的思想家和文學家，被林肯稱為「美國文明之父」，和梭羅（Thoreau，1817 － 1862 年）均為超驗主義理論家，著有《自然論》、《偉人論》、《歷史哲學》、《入世與出世》等。

Chapter VIII

The Reflective Mood

The exercise of concentrating the mind (to which at least half an hour a day should be given) is a mere preliminary, like scales on the piano. Having acquired power over that most unruly member of one's complex organism, one has naturally to put it to the yoke. Useless to possess an obedient mind unless one profits to the furthest possible degree by its obedience. A prolonged primary course of study is indicated.

Now as to what this course of study should be there cannot be any question; there never has been any question. All the sensible people of all ages are agreed upon it. And it is not literature, nor is it any other art, nor is it history, nor is it any science. It is the study of one's self. Man, know thyself. These words are so hackneyed that verily I blush to write them. Yet they must be written, for they need to be written. (I take back my blush, being ashamed of it.) Man, know thyself. I say it out loud. The phrase is one of those phrases with which everyone is familiar, of which everyone acknowledges the value, and which only the most sagacious put into practice. I don't know why. I am entirely convinced that what is more than anything else lacking in the life of the average well-intentioned man of to-day is the reflective mood.

We do not reflect. I mean that we do not reflect upon genuinely important things; upon the problem of our happiness, upon the main direction in which we are going, upon what life is giving to us, upon the share which reason has (or has not) in determining our actions, and upon the relation between our principles and our conduct.

And yet you are in search of happiness, are you not? Have you discovered it?

The chances are that you have not. The chances are that you have already come to believe that happiness is unattainable. But men have attained it. And they have attained it by realising that happiness does not spring from the procuring of physical or mental pleasure, but from the development of reason and the adjustment of conduct to principles.

I suppose that you will not have the audacity to deny this. And if you admit it, and still devote no part of your day to the deliberate consideration of your reason, principles, and conduct, you admit also that while striving for a certain thing you are regularly leaving undone the one act which is necessary to the attainment of that thing.

Now, shall I blush, or will you?

Do not fear that I mean to thrust certain principles upon your attention. I care not (in this place) what your principles are. Your principles may induce you to believe in the righteousness of burglary. I don't mind. All I urge is that a life in which conduct does not fairly well accord with principles is a silly life; and that conduct can only be made to accord with principles by means of daily examination, reflection, and resolution. What leads to the permanent sorrow-fulness of burglars is that their principles are contrary to burglary. If they genuinely believed in the moral excellence of burglary, penal servitude would simply mean so many happy years for them; all martyrs are happy years for them; all martyrs are happy, because their conduct and their principles agree.

As for reason (which makes conduct, and is not unconnected with the making of principles), it plays a far smaller part in our lives than we fancy. We are supposed to be reasonable but we are much more instinctive than reasonable. And the less we reflect, the less reasonable we shall be. The next time you get cross with the waiter because your steak is over-cooked, ask reason to step into the cabinet-room of your mind, and consult her. She will probably tell you that the waiter did not cook the steak, and had no control over the cooking of the steak; and that even if he alone was to blame, you accomplished nothing good by getting cross; you merely lost your dignity, looked a fool in the eyes of sensible men, and soured the waiter, while producing no effect whatever on the steak.

The result of this consultation with reason (for which she makes no charge) will be that when once more your steak is over-cooked you will treat the waiter as a fellow-creature, remain quite calm in a kindly spirit, and politely insist on having a fresh steak. The gain will be obvious and solid.

In the formation or modification of principles, and the practice of conduct, much help can be derived from printed books (issued at sixpence each and upwards). I mentioned in my last chapter Marcus Aurelius and Epictetus. Certain even more widely known works will occur at once to the memory. I may also mention Pascal, La Bruyere, and Emerson. For myself, you do not catch me travelling without my Marcus

Aurelius. Yes, books are valuable. But not reading of books will take the place of a daily, candid, honest examination of what one has recently done, and what one is about to do- -of a steady looking at one's self in the face (disconcerting though the sight may be).

When shall this important business be accomplished? The solitude of the evening journey home appears to me to be suitable for it. A reflective mood naturally follows the exertion of having earned the day's living. Of course if, instead of attending to an elementary and profoundly important duty, you prefer to read the paper (which you might just as well read while waiting for your dinner) I have nothing to say. But attend to it at some time of the day you must. I now come to the evening hours.

第九章
策略八：用文藝樂活人生

無須因為不會彈奏《少女的祈禱》，
就覺得自己不可能成為忠實的樂迷；
就算你只知道樂團中充滿各式各樣的樂器，
它們可以發出各種迷人又不同的聲音，這樣就足夠了。

　　許多人在下班回家後，到上床睡覺前的這段時光裡，通常都是無所事事地在家閒晃。若要他們認真去想點事來充實自我，他們就會望望都快被灰塵淹沒的書架，然後老實地承認：「抱歉！我對閱讀真是一點興趣也沒有。」顯然，他們在生活上有了嚴重的問題。

　　大家都知道，書籍可以幫助我們學習許多事物，透過書的引領，我們可以降低瞎子摸象的無助感，快速地掌握事情的本質，進而解決疑惑。就像當我們想要學會打牌的技巧或是划船的步驟，都可以藉由書籍的說明習得正確的方法。

　　不喜歡閱讀的人，其實是沒有找到自己有興趣的書來閱讀，而非他們以為的——與書天生「無緣」。有很多人以為愛看書的人，其文學素養一定很高，其實書與文學，並非是劃上等號。如果你不擅長研讀文學，何不選擇非文學的書來看，無論如何都是培養閱讀的開始。要知道，即使是一本料理漫畫，它也可以發揮你意想不到的作用。

　　因此就算你不喜歡文學作品，也無須覺得自己的才華低劣，更不用為此感到無地自容。不管你是否讀過梅雷迪思（Meredith，註1）的文章，或搞不清楚斯蒂芬・菲利普斯（Stephen Phillips，註2）是不是一個詩人，其實都沒多大關係。

雖然有些文學家認為，現代人的文學素養普遍極為低落，只因為他們連古詩詞或古文都認識不了幾篇，但是我倒很想請教他們，難道他們就能對柴可夫斯基（Tchaikovsky，註3）所創作的《悲愴交響曲》（Pathetique，註4）瞭若指掌嗎？正所謂「隔行如隔山」，凡是過於自滿於自己世界的人，總是無法體會別人生活中的美好，其實他們的日子也過得很寂寞吧！

在生活中除了文學及閱讀外，還有許多活動都可以陶冶我們的心靈，諸如音樂……等。就在我的腦海裡盤旋起《悲愴交響曲》的樂音時，我突然回想起盛夏的那一場露天音樂會，A君也出席了。

當臺上傳來的美妙樂聲令所有人都如癡如醉之際，A君卻默然地抽起菸來。究竟發生什麼事呢？A君表示，自己根本是個音樂低能兒，他連吉他都搞不清楚幾弦，鋼琴和小提琴就更別說了。雖然聽過幾場音樂會，但總懷疑自己是否能夠消化？我感受到他的自信在動搖。

別再質疑了！此刻，你和你的朋友齊聚一堂，身處在這個人山人海的音樂會場，指揮和樂團為了你們，絞盡腦汁地規劃出一場最完美的演出。你對音樂的鑑賞能力，若與初次聽音樂會時來相比，著實又更上一層樓。要知道，任何人都有與生俱來的欣賞音樂的能力。

幾個月來，你每個星期都花上幾天晚上來聽管絃樂會，無須因為自己不會彈奏《少女的祈禱》（Modlitwa Dziewicy Op. 4，註5），就覺得自己不可能成為忠實的樂迷；更何況，你已經瞭解樂團組成的基本常識，就算你現在只認知到樂團中充滿各式各樣的樂器，它們可以發出各種迷人又不同的聲音，這樣就足夠了。畢竟，你尚未有過這方面的專業訓練，你的耳朵也還未經過開發。

我知道，你非常喜歡聽C小調第五交響曲《命運交響曲》（Symphony No. 5，註6）。雖然你可能無法說

出，樂首主旋律所使用的樂器，但它的旋律卻使你為之
嚮往，直到今日你仍舊念念不忘。我記得，你曾努力地
想要向一位女士推薦它，即使說來說去都只能繞著創作
者貝多芬，但它的確是人人稱頌的世界名曲。

　　如果你希望能多瞭解音樂，我會建議你去看看克比
爾（Krehbiel，註7）的《音樂鑑賞》一書，它可以教你
比你想到的更多。這本書並不難買到，而且它的價格比
聽一場付費音樂會要少得許多。加上它附有許多管絃樂
器的圖片及解說，這會使你以後聽音樂會的功力大增。
從此以後，你聽到的會是所有樂器總合的和諧樂聲，而
再也不是莫衷一是的陌生人。

　　當音樂響起之際，你開始聽出各種樂器的差異性、
不同的音色、演奏的難度，甚至連演出價碼，你也略知
一二。現在的你已經可以凝神傾聽，並判斷樂團及指揮
的表現。你再也不會像過去，不是無聊地猛打瞌睡，就
是只能專注地等著布景會不會變化。

　　對音樂，你已經有了一套屬於自己的認知系統。當
你想要更進一步知道某種形式的音樂，或是某位作曲者

的作品時，你會利用每週三個晚上的時間，來做更深入的研讀。此外因為你鑑賞音樂的能力提升，你也開始會判斷音樂會的性質及水準，是否符合你的需求。

一年後，就算你還是不會以鋼琴彈奏或變調彈奏《少女的祈禱》，那也不是什麼丟臉的事。重要的是，你已經漸漸培養出對音樂欣賞的興趣，你不再是個音樂低能兒。

相對的，我們可以將學習音樂欣賞的方式，比照到其他藝術方面。我們便可以考慮閱讀如克萊蒙‧威特（Clermont Witt，註8）的《畫作鑑賞》及拉塞爾‧斯特吉斯（Russell Sturgis，註9）《建築物鑑賞》……等。總之，各類藝術鑑賞的書可說是琳琅滿目，正等著你去閱讀。

最後，在聽完這麼多例子後，如果還是無法激起你嘗試欣賞藝術的興趣……「是的，我就是討厭藝術！」

是的，我也會尊重你的想法。然後在下一章中，我們將來研究你為何會有這種想法，以及你的決定會對生活產生什麼影響。

註 1 梅雷迪思（Meredith，1818 － 1873 年），英國的古典派學者。

註 2 斯帝芬·菲利普斯（Stephen Phillips，1864 － 1915 年），英國詩人兼戲劇家，職業生涯早期就享有高知名度。

註 3 柴可夫斯基（Pyotr Ilyich Tchaikovsky，1840 － 1893 年）是俄國著名的作曲家，他的作品旋律優美動人，著名的作品有《天鵝湖》、《胡桃鉗》等芭蕾舞劇和《1812 序曲》、《悲愴交響曲》等，都是深受歡迎的作品。

註 4 《悲愴交響曲》（Pathetique），柴可夫斯基於 1893 年完成的最後一首音樂作品，也是他的六首交響曲中最傑出的一首，被列為「世界六大通俗交響曲」之一。這首交響曲名為《悲愴》，是柴可夫斯基一生的寫照，由於他的性格一向憂鬱，其厭世的情緒使他自知不久人世，便把情緒都傾吐於此作品上。

註 5 《少女的祈禱》（Modlitwa Dziewicy Op. 4）：由波蘭女作曲家特克拉·巴達捷夫斯卡 - 巴拉諾夫斯卡於 1859 年發表的傳世名作。

一天始於下班後：

妥善運用你的待機時間

註 6 C 小調第五交響曲《命運交響曲》（Symphony No. 5）：著名
音樂家貝多芬於 1804 至 1808 年間創作的四樂章交響曲。該作品
可說是史上所有交響曲中最受歡迎的一首，其演出次數亦居所有交
響曲之冠。

註 7 克比爾（Henry Edward Krehbiel，1854 － 1923 年），美國音
樂評論家和音樂學家，著有多本音樂評論的書籍，如：《音樂鑑
賞》、《歌劇集》等。

註 8 克萊蒙‧威特（Clermont Witt，1872 － 1952 年），英國藝術
史學家，是倫敦科陶爾德藝術學院的聯合創始人之一。

註 9 拉塞爾‧斯特吉斯（Russell Sturgis，1836 － 1909 年），美
國著名建築師兼建築史學家。

Chapter IX

Interest in the Arts

Many people pursue a regular and uninterrupted course of idleness in the evenings because they think that there is no alternative to idleness but the study of literature; and they do not happen to have a taste for literature. This is a great mistake.

Of course it is impossible, or at any rate very difficult, properly to study anything whatever without the aid of printed books. But if you desire to understand the deeper depths of bridge or of boat-sailing you would not be deterred by your lack of interest in literature from reading the best books on bridge or boat-sailing. We must, therefore, distinguish between literature, and books treating of subjects not literary. I shall come to literature in due course.

Let me now remark to those who have never read Meredith, and who are capable of being unmoved by a discussion as to whether Mr. Stephen Phillips is or is not a true poet, that they are perfectly within their rights. It is not a crime not to love literature. It is not a sign of imbecility. The mandarins of literature will order out to instant execution the unfortunate individual who does not comprehend, say, the influence of Wordsworth on Tennyson. But that is only their impudence. Where would they be, I wonder, if requested to explain the influences that went to make Tschaikowsky's "Pathetic Symphony"?

There are enormous fields of knowledge quite outside literature which will yield magnificent results to cultivators. For example (since I have just mentioned the most popular piece of high-class music in England to-day), I am reminded that the Promenade Concerts begin in August. You go to them. You smoke your cigar or cigarette (and I regret to say that you strike your matches during the soft bars of the

"Lohengrin" overture), and you enjoy the music. But you say you cannot play the piano or the fiddle, or even the banjo; that you know nothing of music.

What does that matter? That you have a genuine taste for music is proved by the fact that, in order to fill his hall with you and your peers, the conductor is obliged to provide programmes from which bad music is almost entirely excluded (a change from the old Covent Garden days!).

Now surely your inability to perform "The Maiden's Prayer" on a piano need not prevent you from making yourself familiar with the construction of the orchestra to which you listen a couple of nights a week during a couple of months! As things are, you probably think of the orchestra as a heterogeneous mass of instruments producing a confused agreeable mass of sound. You do not listen for details because you have never trained your ears to listen to details.

If you were asked to name the instruments which play

the great theme at the beginning of the C minor symphony
you could not name them for your life's sake. Yet you admire
the C minor symphony. It has thrilled you. It will thrill
you again. You have even talked about it, in an expansive
mood, to that lady-you know whom I mean. And all you
can positively state about the C minor symphony is that
Beethoven composed it and that it is a "jolly fine thing."

Now, if you have read, say, Mr. Krehbiel's "How to
Listen to Music" (which can be got at any bookseller's for
less than the price of a stall at the Alhambra, and which
contains photographs of all the orchestral instruments and
plans of the arrangement of orchestras) you would next go
to a promenade concert with an astonishing intensification
of interest in it. Instead of a confused mass, the orchestra
would appear to you as what it is--a marvellously balanced
organism whose various groups of members each have a
different and an indispensable function. You would spy out
the instruments, and listen for their respective sounds. You

would know the gulf that separates a French horn from an English horn, and you would perceive why a player of the hautboy gets higher wages than a fiddler, though the fiddle is the more difficult instrument. You would *live* at a promenade concert, whereas previously you had merely existed there in a state of beatific coma, like a baby gazing at a bright object.

The foundations of a genuine, systematic knowledge of music might be laid. You might specialise your inquiries either on a particular form of music (such as the symphony), or on the works of a particular composer. At the end of a year of forty-eight weeks of three brief evenings each, combined with a study of programmes and attendances at concerts chosen out of your increasing knowledge, you would really know something about music, even though you were as far off as ever from jangling "The Maiden's Prayer" on the piano.

"But I hate music!" you say. My dear sir, I respect you.

What applies to music applies to the other arts. I might mention Mr. Clermont Witt's "How to Look at Pictures," or Mr. Russell Sturgis's "How to Judge Architecture," as beginnings (merely beginnings) of systematic vitalising knowledge in other arts, the materials for whose study abound in London.

"I hate all the arts!" you say. My dear sir, I respect you more and more. I will deal with your case next, before coming to literature.

第十章
策略九：生活從想法開始

事情真的沒有你想像中的困難。

你只要知道推斷日常事物的方法，同理可證，

人生中再複雜的事一樣可解。

重點是你願意放下偏見及固執，重新完整地省視事物

嗎？

雖然說品味藝術是件很好的事，但並非是人生中最為重要的事。

想要透徹地瞭解生命，就要學會深入觀察事物的因果關係，亦即，其不斷演化的過程。更進一步說，也就是要針對人類文明變化的歷史，做更細微的研究。

俗話說：「每件事發生都有其道理。」是的，有因就有果，有果必有因。若想真正地認識各種事物，我們必然要去追求其發生的原因和變化的因素，才能確信其導致的結果。

當一個人不再盲目地只相信眼見為憑，而懂得凡事都要追本溯源時，他的心思會越來越細膩，思慮會越來越清晰，胸襟也會更加廣大，而他的世界必然會比一般人，更為遼闊。

舉例來說，當你的手錶失竊了，當下想必你會十分憤怒及沮喪，但若靜下心來想想，這位小偷為何要竊走你的手錶呢？是惡性不改？臨時起意？還是因為家境或不為人知的悲慘原因導致的呢？雖然想到了千百種原因，可能也無法喚回手錶，但你卻漸漸瞭然於胸。總之，

那支手錶不見必然有其原因，但你可以改變的，是學會
更小心謹慎地保管物品。於是，你終於可以豁然開朗地
再去買支新錶了。

　　這正如當我們遭逢生命中的逆境時，我們總是先本
能地怨天尤人一番，甚至因此就被打擊及創傷的陰霾所
淹沒。可是若平心靜氣地回想前因後果，重新檢視自己
的信念及做法，就會發現，原來事情的發生果然不是沒
道理的。如此，自己便能夠勇於面對事實，並脫離愁雲
慘霧的困境。

　　許多人都認為，人性是這世上最深不可測及令人感
到害怕的事，但事實果真是如此嗎？隨著我們的成長經
歷，思想也會越來越成熟。偶爾自己難免也會犯了人性
的錯，但任何事物都有其一體兩面，有惡必有善，重要
的是我們是否有正確的信念，而非一味歸咎於「他人」
的人性。若再回想起來，自己最初對人性戒慎恐懼的模
樣，是否也會忍不住莞爾一笑？

　　因此想要避免對人性不必要的恐慌症，化解對生命
的無力感及困惑，我們就要加強自己剖析事物的因果由

來的能力，以及研究改善境遇的解決之道。如此一來，不但可以減輕人生中所遭受到的痛苦及挫敗，也因為思考而增加了許多生活上的趣味性。

一個能深悟生命無常及宇宙變化道理的人，與一個只懂得生存之道的人，他們最大的差別在於：**瞭解思考的重要性**。就好像同時讓兩者去看海，前者因為知道大自然生態變化的各種原因，所以他可以看見呈現出不同的海洋世界的風貌，正如他明瞭人生無常，變幻莫測，所以一定要把握當下，珍惜擁有；而後者所看到的海，永遠就是一望無際的遼闊，亦如他的生活千篇一律，乏善可陳。

要學習分析事物的因果關係，其實生活中有許多地方就可以印證。例如有人傳說 A 地區的房價又上漲了，雖然這個消息令許多荷包原本就癟癟的小老百姓為之扼腕，但就某一方面來說，假若現在我們是該地區的因果關係研究員，那麼，你將要如何證明其真的上漲了呢？

為了要充分掌握線索，你決定深入 A 地明察暗訪。於是，你混入該地最多人用餐的餐廳調查消息來源的可靠性，結果，你從許多在地居民及上班族處得知：原來，該區政府為了繁榮發展，特別發放給該區或進入該區工作的人交通補助費，這樣一來便使得Ａ區的房屋供不應求，房價當然隨之水漲船高了。

　　或許有人會忍不住想揶揄：「不會吧！想要瞭解人生的因果，怎麼可能會像探查Ａ區房價這事一樣這麼簡單呢？」是的，事情真的沒有你想像中的困難。你只要知道推斷一件日常事物的方法，同理可證，人生中再複雜的事一樣可解。重點是你願意回歸初心，放下偏見及固執，重新地、重頭地、完整地省視事物一番嗎？

　　這就好比你是一名房屋仲介公司的經紀人，但很不幸的，若非為了生計，說什麼你也不想當個整天要看人臉色的業務員，但為了保住飯碗，你還是每天忍氣吞聲地向客人低聲下氣地解說。事實上，你真是厭倦極了這份工作。

　　但是，我給你的忠告是：這世上，存在的萬事萬物都有其意義的。

　　好吧！讓我們一起回到你的辦公室，來檢視你的工作，你將發現，原來這個職業蘊藏著許多的知識及樂趣啊！首先，關於Ａ地房價上漲的另一個傳說，原來是因為Ｂ區的勞動法規調整，導致企業的人事成本增加，物價也跟著高漲，使得許多人口外移到Ａ區找房子。有了這個新發現，難道你還不清楚自己的工作需要許多專業分析嗎？這份職業其實很有趣的！

我們再深入地回想，假如你願意讓自己挑戰更高難度的問題，例如你認為Ｂ區人口會外移到Ａ區，其實並非因為物價或是交通費補助，而是你發現Ｂ區頒布的新建築法規對成屋的容積將造成的結果影響所致；又或者，你願意每天下班後多花一個半小時，更加深入地研究Ａ、Ｂ地區的房產及人口變化。經過一段時間後，你將會看見自己的專業能力不但提升，你對工作又重新燃起十足的動力，更重要的，是你的生活將煥然一新。

如果一個房屋仲介的經紀人案例，還不足以讓你認同我的論點，那我們再來假設，你是一位任職於投顧公司的理財專員吧！請別告訴我，你還未曾認真研讀過任何世界知名企業家的人生傳記，不要以為那只是關於他的生命歷史，和你的投資專業沒有任何關係。問題就在此處，如果你能每天多花個二十分鐘，研讀那些商業鉅子的成功之道，或許你就可以早一天成為財經界的投資天王（天后）。

再如，你是個不喜歡被「拘禁」在城市中的人，奧妙的大自然反倒是最吸引你的地方。因為寬闊無邊的視野，總是讓你苦悶的心獲得解放。雖然受制於生活條件，你無法常常讓自己「野放」，那何不嘗試在自己的周遭環境中，多製造一些綠意的效果呢？或許在一個不預期的日子裡，你會在陽臺上小小的盆栽中發現難得一見的蝴蝶，你甚至可以進而研究栽種植物的生態變化，說不定還意外地成為你獨有的發表學說。但是，**總歸你真正的、最大的收穫卻是藉由興趣的培養，而改變了對人生的看法。**

總之，投資自我的方法有千百萬種，更絕非只有閱讀一途。只要你願意每天不間斷地花上固定的時數，不論是了自己的工作、生活、家庭，甚至是樂趣，深入而廣大地研究它，不僅可以使你在職場如魚得水，同時也能有效地面對生活，盡嘗人生中的許多樂趣。

在我們的生活中，許多時候我們總是深深地受著習慣及環境的影響，想讓人生豐富精彩，這其中當然需要旺盛的好奇心及身心平衡的滿足感。對生活中的萬事萬物都感到新奇有趣，想要去開發，想要去研究，充滿如初生之犢的勇氣驅動著我們往更高、更遠的境界前進。最後，終於能全然地接納自己的人生，理解生命的可貴之處，我們的所有欲望也將得到完全的滿足。

在上一章的末段中，我曾承諾要讓你深入地瞭解，為何你會不喜愛藝術的原因，及會使你的生活產生的影響，經過這一章的說明，我想你應該十分瞭解，一切都是你的想法在作怪吧！所幸的是，想法是可以改變的，重要的是你要有改變的意願，至於下一章，我將提供改變的方法讓你參考。

Chapter X

Nothing in Life Is Humdrum

Art is a great thing. But it is not the greatest. The most important of all perceptions is the continual perception of cause and effect-in other words, the perception of the continuous development of the universe-in still other words, the perception of the course of evolution. When one has thoroughly got imbued into one's head the leading truth that nothing happens without a cause, one grows not only large-minded, but large-hearted.

It is hard to have one's watch stolen, but one reflects that the thief of the watch became a thief from causes of heredity and environment which are as interesting as they are scientifically comprehensible; and one buys another watch, if not with joy, at any rate with a philosophy that makes bitterness impossible. One loses, in the study of cause and effect, that absurd air which so many people have of being always shocked and pained by the curiousness of life. Such people live amid human nature as if human nature were a foreign country full of awful foreign customs. But, having reached maturity, one ought surely to be ashamed of being a stranger in a strange land!

The study of cause and effect, while it lessens the painfulness of life, adds to life's picturesqueness. The man to whom evolution is but a name looks at the sea as a grandiose, monotonous spectacle, which he can witness in August for three shillings third-class return. The man who is imbued with the idea of development, of continuous cause and effect, perceives in the sea an element which in the day-before-yesterday of geology was vapour, which yesterday was boiling, and which to-morrow will inevitably be ice.

He perceives that a liquid is merely something on its way to be solid, and he is penetrated by a sense of the tremendous, changeful picturesqueness of life. Nothing will afford a more durable satisfaction than the constantly cultivated appreciation of this. It is the end of all science.

Cause and effect are to be found everywhere. Rents went up in Shepherd's Bush. It was painful and shocking that rents should go up in Shepherd's Bush. But to a certain point we are all scientific students of cause and effect, and there was not a clerk lunching at a Lyons Restaurant who did not scientifically put two and two together and see in the (once) Two-penny Tube the cause of an excessive demand for wigwams in Shepherd's Bush, and in the excessive demand for wigwams the cause of the increase in the price of wigwams.

"Simple!" you say, disdainfully. Everything-the whole complex movement of the universe-is as simple as that-when you can sufficiently put two and two together. And, my dear sir, perhaps you happen to be an estate agent's clerk, and you hate the arts, and you want to foster your immortal soul, and you can't be interested in your business because it's so humdrum.

Nothing is humdrum.

The tremendous, changeful picturesqueness of life is marvellously shown in an estate agent's office. What! There was a block of traffic in Oxford Street; to avoid the block people actually began to travel under the cellars and drains, and the result was a rise of rents in Shepherd's Bush! And you say that isn't picturesque! Suppose you were to study, in this spirit, the property question in London for an hour and a half every other evening. Would it not give zest to your business, and transform your whole life?

You would arrive at more difficult problems. And you would be able to tell us why, as the natural result of cause and effect, the longest straight street in London is about a yard and a half in length, while the longest absolutely straight street in Paris extends for miles. I think you will admit that in an estate agent's clerk I have not chosen an example that specially favours my theories.

You are a bank clerk, and you have not read that breathless romance (disguised as a scientific study), Walter Bagehot's "Lombard Street"? Ah, my dear sir, if you had begun with that, and followed it up for ninety minutes every other evening, how enthralling your business would be to you, and how much more clearly you would understand human nature.

You are "penned in town," but you love excursions to the country and the observation of wild life-certainly a heart-enlarging diversion. Why don't you walk out of your house door, in your slippers, to the nearest gas lamp of a night with a butterfly net, and observe the wild life of common and rare moths that is beating about it, and co-ordinate the knowledge thus obtained and build a superstructure on it, and at last get to know something about something?

You need not be devoted to the arts, not to literature, in order to live fully.

The whole field of daily habit and scene is waiting to satisfy that curiosity which means life, and the satisfaction of which means an understanding heart.

I promised to deal with your case, O man who hates art and literature, and I have dealt with it. I now come to the case of the person, happily very common, who does "like reading."

第十一章
策略十：從「悅讀」與世界接軌

文學的形式有許多種，詩的表現可謂是最高難度，
也最能使人感到欣喜若狂的。

詩歌，雖然最讓人腸枯思竭，它也讓人獲得最多的智
慧。

請容我再一次呼籲：思考，就從讀詩開始。

　　如果你已經決定，要真正省思思考的重要性，這真是一件值得慶賀的事。但是，假若你選擇的方式，是每週抽出三個晚上各一個半小時的時間，來研討查爾斯·狄更斯（Charles Dickens，註 1）的小說，那我可否請你再多三思一下。因為閱讀大部分的小說，對思考力的訓練而言，實在有限。

　　我的意思並非指小說不需要思考，真正的問題是：我們不需要浪費時間去閱讀劣質的作品。另一方面來說，一部好的小說作品，讀者根本不須費吹灰之力就能輕鬆看完，例如梅雷迪思的小說，或許有些小小的瑕疵，但整體讀來，還是令人意猶未盡。

　　對讀者來說，閱讀到優質的小說，就如同搭乘一葉輕舟飛瀑而下的暢快感，當抵達終點站時，或許會因為驚喜而有些目眩神迷，卻絕對不會令人感到疲憊不堪。除此之外，更別說是看到驚為天人的小說，那就更如餘音繞梁三日，令人久久無法忘懷。

　　但是在訓練思考力時，卻不適合這樣「一路順暢」的狀況。我們必須經常刺激腦子，時時刻刻保持著對立和質疑的心態，及尋求真理的毅力。因此，可能會感到既辛苦又痛苦。由於破壞與建設在心中反覆地爭鬥中，所以才會逐漸使我們思考的層次不斷往前邁進。回想一下，當你在讀《安娜·卡列尼娜》（Anna Karenina，註2）時，需要如此「費盡心機」嗎？

在所有的文學中，詩的表現形式可謂是最高難度的。這也是為何讀詩比看小說要更加「用心」的原因。詩歌，雖然是最讓人腸枯思竭、絞盡腦汁的，但是它也讓人獲得最多的智慧。雖然文學的形式有許多種，但詩歌是其中最能使人感到振奮及欣喜若狂的。只可惜大部分的人，都未曾體驗過詩歌的好處。

我曾經做過一份調查，我問朋友說：「如果要你在下列兩件事擇一去做，你會如何選擇？第一、閱讀《失樂園》（Paradise Lost，註3）；第二、在大街上扮演乞丐討錢？」結果，所有的朋友都毫不猶豫地說：「第二。」雖然如此，請容我再一次苦口婆心地呼籲：思考，就從讀詩開始。

或許我們討論到這裡，有些人還是認為，詩歌是文學中最折磨人的刑具；是抽象得像一頭發了瘋的大象；
是無聊到可以殺人無數的槍械，所以你對詩歌的態度仍
然是「敬而遠之」。那麼，我想推薦你閱讀一部關於詩
歌基礎理論的名著——哈茲利特（Hazlitt，註4）的《詩
歌概論》。至少，你不會對詩歌再有太多的偏見了。

　　我的朋友在接受我的建議後，起初也是壓抑住不耐
煩的心情，終於看完哈茲利特的《詩歌概論》後，沒想
到過了一陣子，他竟然在聚會時開始有了朗誦詩歌的想
法。若是你在閱讀完後也受到激勵，想要進一步地瞭解
詩歌，敘事詩是個不錯的選擇。

　　英國有一名女作家伊莉莎白・巴雷特・白朗寧
（Elizabeth Barrett Browning，註5），她用詩體寫了一
本書《奧蘿拉・莉》（Aurora Leigh，註6）。我認為
這本書的成就，不但可說超越女性作家喬治・艾略特
（George Eliot，註7）及勃朗特三姊妹（The Bronte，註
8），甚至可說是比珍・奧斯汀（Jane Austen，註9）的
任何一部作品，都更為出色。

　　在歌頌《奧蘿拉・莉》的同時，我想請你抱著一種

「至死方休」的精神研讀，並暫時將它當作一本優秀的小說。我相信，大部分的人在看完這本著作後，一定會想不通自己為何曾經那麼討厭詩歌呢？因為詩中的一切實在太不可思議了。

如果在讀過哈茲利特的論理後，即使照著他的引導進入了詩的國度，但最後，你仍然一刻都不想多停留的話……

雖然有些遺憾，但我們也不要再浪費時間了，那就朝歷史及哲學類的書籍出發吧！就如《羅馬帝國衰亡史》（The History of the Decline and Fall of the Roman Empire，註 10）雖不能與《失樂園》相提並論，但比起許多書來，它還是一本相當有分量的好書。

或許有人會說，讀詩是需要一些天分吧！但我卻認為，即使是資質再平庸的人，只要經過一年不間斷的閱讀，一定可以理解所有歷史及哲學的論述。因為這方面的著作，不同於需要想像或艱澀難懂的詩歌，它們是非常容易就讓人清楚明白的。

基本上，我不會再建議你要讀哪些歷史或哲學的書，因為那樣的嘗試並沒有太大的意義。但是有兩項原則請你務必要注意：

第一、請確定研讀的目標及範圍。

不管是某個期間，或是某項主題，甚至是某位作者，亦即，你可以研究法國大革命的那一段期間的歷史，或是世界鐵路史的淵源和演變，又或是作家約翰‧慈濟（John Keats，註 11）的作品。

總之，利用一段固定的時間，集中注意力在自己設定的課題上，並設法解決心中的疑問。畢竟若能經由努力不懈的研究，成為某門專科的達人，也是人生中意外的收穫。

第二、閱讀時，請同時思考。

有些人雖然閱讀成癮，但是閱讀對他的人生卻猶如麵包上的奶油，吃過後毫無痕跡。即使他們看過的書堆積如山，看起書來也是行雲流水的輕易，一年到頭始終保持優雅的讀書人姿態，逢人便誇耀自己閱書無數，到了書店也能如數家珍，但反觀他的生活方式，依然沒有太大的進展或變化。書，顯然對他的人生沒太大的用處。

　　既然花了二十分鐘的時間看書，那最少也要用四十五分鐘反覆思考。若是無法從書本中想出些道理，那就無須浪費時間閱讀了。此外，一開始看書，可能會覺得力不從心，此時不須操之過急，逐步逐量增加即可。

　　閱讀時，不要一心只想著達到目標。認真地埋首其中，讓自己完全沉浸於書的世界。往往就在最不經意的時候，你會赫然地發現，自己的人生似乎到了一個美麗新境界。

註 1 查爾斯‧狄更斯（Charles Dickens，1812 － 1870 年），維多利亞時代的著名英國作家。童年生活貧困，長大後成為記者，並陸續出版多部經典知名小說。他的小說大多為描寫英國中下層社會的故事和人物，如：《雙城記》、《孤雛淚》、《塊肉餘生錄》、《聖誕鬼異》等。

註 2 《安娜‧卡列尼娜》（Anna Karenina），由俄國小說家托爾斯泰（Leo Tolstoy，1828 － 1910 年）所寫。故事描述一位貴族夫人挑戰虛偽的上流社會，勇敢追求真愛的故事。

註 3 《失樂園》（Paradise Lost），17 世紀英國詩人約翰‧彌爾頓以史詩的形式寫成的作品。全詩共分 12 書，內容立基於舊約聖經，內容大意為：人類的祖先亞當和夏娃，因為違背了上帝的旨意，接觸惡魔撒旦並偷嘗伊甸園中的禁果，不僅使人類遭受處罰，並被逐出樂園的故事。

註 4 威廉‧哈茲利特（William Hazlitt，1778 － 1830 年），英國散文家、戲劇和文學評論家、畫家、社會評論家和哲學家，被認為是英語史上最偉大的評論家和散文學家之一。

註 5 伊莉莎白‧巴雷特‧白朗寧（Elizabeth Barrett Browning，1806 － 1861 年），是英國維多利亞時代最著名的詩人之一。作品涉及廣泛的議題和思想，影響多位同一時期的人物。著有《葡語十四行詩集》等多篇詩作和小說。

註 6 《奧蘿拉‧莉》（Aurora Leigh）：1856 年問世，伊莉莎白稱此詩為「詩歌中的小說」，部分可說為伊莉莎白之自傳，內容主要闡述當代婦女角色地位之評論，極具社會挑戰性，被約翰‧羅斯金（John Roskin）稱為 19 世紀最偉大的長詩。

註 7 喬治‧艾略特（George Eliot，1819 － 1880 年），英國小說家，

著有《佛羅斯河畔上的磨坊》、《米德爾馬契》、《教區生活場景》
和《掀開面紗》等書。

註8 勃朗特三姊妹（The Bronte）：著名的英國文壇三姊妹。夏綠蒂．
勃朗特（Charlotte Bronte，1816 － 1855 年），代表作：《簡愛》；
艾蜜莉．勃朗特（Emily Bronte， 1818 － 1848 年），代表作：《咆
哮山莊》；安妮．勃朗特（Anne Bronte，1820 － 1849 年），代表
作：《荒野莊園的房客》。

註9 珍．奧斯汀（Jane Austen，1775 － 1817 年），著名英國小說
家，重要著作有：《傲慢與偏見》、《理性與感性》、《曼斯菲爾
德莊園》、《愛瑪》、《諾桑覺寺》和《勸導》等多部小說。

註10 《羅馬帝國衰亡史》（The History of the Decline and Fall of
the Roman Empire）：英國歷史學家愛德華．吉朋（Edward Gibbon，
1737 － 1794 年）歷經二十餘年的心血所寫，為西洋史學巨著，全
書上下縱橫一千三百年，共分六大冊。

註11 約翰．慈濟（John Keats，1795 － 1821 年），出生於 18 世
紀末的倫敦，傑出的英詩作家之一，也是浪漫派的主要成員。在病
中寫出了大量優秀作品，包括《聖艾格尼絲之夜》、《夜鶯頌》和
《致秋天》等。

Chapter XI

Serious Reading

Novels are excluded from "serious reading," so that the man who, bent on self-improvement, has been deciding to devote ninety minutes three times a week to a complete study of the works of Charles Dickens will be well advised to alter his plans. The reason is not that novels are not serious--some of the great literature of the world is in the form of prose fiction--the reason is that bad novels ought not to be read, and that good novels never demand any appreciable mental application on the part of the reader. It is only the bad parts of Meredith's novels that are difficult. A good novel rushes you forward like a skiff down a stream, and you arrive at the end, perhaps breathless, but unexhausted. The best novels involve the least strain. Now in the cultivation of the mind one of the most important factors is precisely the feeling of strain, of difficulty, of a task which one part of you is anxious to achieve and another part of you is anxious to shirk; and that f eeling cannot be got in facing a novel. You do not set your teeth in order to read "Anna Karenina." Therefore, though you should read novels, you should not read them in those ninety minutes.

Imaginative poetry produces a far greater mental strain than novels. It produces probably the severest strain of any form of literature. It is the highest form of literature. It yields the highest form of pleasure, and teaches the highest form of wisdom. In a word, there is nothing to compare with it. I say this with sad consciousness of the fact that the majority of people do not read poetry.

I am persuaded that many excellent persons, if they were confronted with the alternatives of reading "Paradise Lost" and going round Trafalgar Square at noonday on their knees in sack-cloth, would choose the ordeal of public ridicule. Still, I will never cease advising my friends and enemies to read poetry before anything.

If poetry is what is called "a sealed book" to you, begin by reading Hazlitt's famous essay on the nature of "poetry in general." It is the best thing of its kind in English, and no one who has read it can possibly be under the misapprehension that poetry is a mediaeval torture, or a mad elephant, or a gun that will go off by itself and kill at forty paces. Indeed, it is difficult to imagine the mental state of the man who, after reading Hazlitt's essay, is not urgently desirous of reading some poetry before his next meal. If the essay so inspires you I would suggest that you make a commencement with purely narrative poetry.

There is an infinitely finer English novel, written by a woman, than anything by George Eliot or the Brontes, or even Jane Austen, which perhaps you have not read. Its title is "Aurora Leigh," and its author E.B. Browning. It happens to be written in verse, and to contain a considerable amount of genuinely fine poetry. Decide to read that book through, even if you die for it. Forget that it is fine poetry. Read it simply for the story and the social ideas. And when you have done, ask yourself honestly whether you still dislike poetry. I have known more than one person to whom "Aurora Leigh" has been the means of proving that in assuming they hated poetry they were entirely mistaken.

Of course, if, after Hazlitt, and such an experiment made in the light of Hazlitt, you are finally assured that there is something in you which is antagonistic to poetry, you must be content with history or philosophy. I shall regret it, yet not inconsolably. "The Decline and Fall" is not to be named in the same day with "Paradise Lost," but it is a vastly pretty thing; and Herbert Spencer's "First Principles" simply laughs at the claims of poetry and refuses to be accepted as aught but the most majestic product of any human mind. I do not suggest that either of these works is suitable for a tyro in mental strains. But I see no reason why any man of average intelligence should not, after a year of continuous reading, be fit to assault the supreme masterpieces of history or philosophy. The great convenience of masterpieces is that they are so astonishingly lucid.

I suggest no particular work as a start. The attempt would be futile in the space of my command. But I have two general suggestions of a certain importance. The first is to define the direction and scope of your efforts. Choose a limited period, or a limited subject, or a single author. Say to yourself: "I will know something about the French Revolution, or the rise of railways, or the works of John Keats." And during a given period, to be settled beforehand, confine yourself to your choice. There is much pleasure to be derived from being a specialist.

The second suggestion is to think as well as to read. I know people who read and read, and for all the good it does them they might just as well cut bread-and-butter. They take to reading as better men take to drink. They fly through the shires of literature on a motor-car, their sole object being motion. They will tell you how many books they have read in a year.

Unless you give at least forty-five minutes to careful, fatiguing reflection (it is an awful bore at first) upon what you are reading, your ninety minutes of a night are chiefly wasted. This means that your pace will be slow.

Never mind.

Forget the goal; think only of the surrounding country; and after a period, perhaps when you least expect it, you will suddenly find yourself in a lovely town on a hill.

第十二章
小心！時間管理的陷阱

在我們學會如何管理好一生的時間之前，
地球依然運轉，
不管最後我們的人生是否圓滿，
地球還是不會停止運作，
這正是我們應該學習的人生態度。

希望截至目前為止，我所提供關於如何充分利用時間的建議，不至於讓讀者覺得是冗長的說教，毫無一點用處。畢竟，我們的人生可不同於一般動物或植物的生存。本書最終的目的，仍是希望藉由十個時間策略，能幫助讀者瞭解並重視，掌握每一天寶貴的時間，進而擁有一個充實快樂的人生。

在這最後一個章節中，我依然要一再地向你提出建言，尤其是我要特別指出，當我們在進行時間管理時，可能會遭遇到的幾項問題：

第一、自己的時間，自己安排。

當有人建議你，要將一切時間用於工作，才是最正確的事，請千萬不要相信他。你的時間必須要由自己掌握，因為他是不可能負責你的人生的。

讓我們先來探討一下，關於那位告訴你「工作勝於一切」的那個人，他究竟是個怎樣的人？首先，他應該是自視甚高，認為自己是很有智慧的人，同時，他也是不可一世的，以為凡人都需要他的教化。但是在旁人心中的他，卻遠非如他所想。

　　事實上，他的傲慢、他的專制、他的無禮，在在都使人對他生厭。他們總是想盡辦法要吸引人家的目光，於是，他自以為是地大放厥詞，胡亂地指揮著那些迷惘的羔羊，錯以為自己成了救贖的主。當有人發現了他的詭計，拆穿了他的謊言，他便會惱羞成怒地說別人的不是。

　　但是，我們更應該引以為戒的是，千萬不要讓自己成了那樣的人。不僅如此，我們還要避免對他人的生活說三道四，更無須指責浪費時間的人。那都是他人的人生，與我們無關。我們真正能做的、該做的，是盡自己的本分，管好自己的人生。

　　所謂：「天行健，君子以自強不息。」要知道，在我們學會如何管理好自己一生的時間之前，地球依然不眠不休地運轉，不管最後我們是否有了圓滿的人生，地球還是不會停止它的運作，這也正是我們應該學習的人生態度。

　　第二、不要將自己變成計畫的奴隸。

　　要想真正地落實計畫，就要規劃出自己能力範圍內

的計畫。為了表示對計畫的尊重，我們要抱著誠懇的心
確實地執行。但這並非表示計畫是神聖不可侵犯的，它
不是神祇，我們更非盲目的信徒，不要將自己畫地自限，
受制於計畫的框架中。

　　大部分的人應該都能瞭解其中的道理，但偏偏就
是有人執迷不悟，將自己和親友都帶入這個痛苦的深淵
中。一位深受其害的妻子就曾向我訴苦：「我真是受不
了了！我的老公簡直就像是娶了計畫一樣。早上八點，
他一定會去遛狗；八點四十五分，就開始讀書。連我想
單獨跟他一起共同做些事，都成了不可能的任務。」從
她哀怨的眼神中，我彷彿看見了那個冥頑不靈的老公和
他們悲慘的婚姻。

　　從另一方面來說，由於計畫只是計畫，如果不切實
執行，它就成了可笑的白日夢。因此，要認真務實地面
對計畫，卻也不用到戒慎恐懼的地步。對於計畫的新手
來說，這一點還需要經過一些體驗才能有所明瞭。

　　第三、不要在倉促間執行計畫。

　　在匆忙間實施的計畫，往往會讓自己陷於下一件事

的焦慮之中。例如上例中的丈夫，當他在清晨八點遛狗的同時，心裡就會開始念著要在八點四十五前趕回家裡讀書。如此一來，他讓自己的生活受限於計畫之中，如同活在監獄中的囚犯。

有時為了改進事情的發展，我們可以試圖改變計畫。由於貪念使然，我們經常會在不覺中訂定太多的計畫，甚至已經超出自己的能力所及。此時，就要懂得節制需求，並減少不必要或緩不濟急的計畫。

但是如同擁有越多知識，求知欲就會更加索求無度，很多人就是喜歡忙得不可開交，即使精疲力盡，他們還是寧可選擇不斷找事來做。畢竟，這樣他們的心裡才有一種踏實感。

當計畫超出你的能力以外，但你又不願改變它時，有一個緩衝的方法是：折衷一部分的時間改做其他的活動或休息。如前例中的丈夫，在完成遛狗任務要回家的途中，可先花個五分鐘的時間，讓自己平靜下來，完全不想任何事情。亦即，讓這五分鐘成為一片空白，就算浪費了也沒關係。

第四、避免在計畫實施之初，就遭受到失敗。

這是執行計畫時最大的危機，在之前的幾章，我就曾一再地提醒大家。失敗會吞噬掉計畫初始的信心與行動，就如同殺害未來會成為大樹的新芽一般。因此絕對要小心預防在計畫起步時就慘遭滑鐵盧的命運，循序漸進地實踐，切勿操之過急。另外，也不要讓它背負沉重的壓力導致執行上的困頓。

既然訂定目標，當然就要努力完成，無論過程會有任何困難與否，我們都要全力以赴。一旦完成任務後，除了目的達成外，最大的禮物是無以言喻的自信。最後，當你計畫要投資你的人生時，**請選擇從自己的興趣開始，其次，一定要與自己的性格相合。**

若想要從瞭解生命為出發點，哲學類的百科全書是個不錯的建議。如果對艱深的哲學語言實在沒有多大興致，而想研究趣味性的話題，諸如街頭叫賣小販的歷史淵源，那就把百科全書暫擱一旁，從你有興趣的地方開始吧！

Chapter XII

Dangers to Avoid

I cannot terminate these hints, often, I fear, too didactic and abrupt, upon the full use of one's time to the great end of living (as distinguished from vegetating) without briefly referring to certain dangers which lie in wait for the sincere aspirant towards life. The first is the terrible danger of becoming that most odious and least supportable of persons--a prig.

Now a prig is a pert fellow who gives himself airs of superior wisdom. A prig is a pompous fool who has gone out for a ceremonial walk, and without knowing it has lost an important part of his attire, namely, his sense of humour. A prig is a tedious individual who, having made a discovery, is so impressed by his discovery that he is capable of being gravely displeased because the entire world is not also impressed by it. Unconsciously to become a prig is an easy and a fatal thing.

Hence, when one sets forth on the enterprise of using all one's time, it is just as well to remember that one's own time, and not other people's time, is the material with which one has to deal; that the earth rolled on pretty comfortably before one began to balance a budget of the hours, and that it will continue to roll on pretty comfortably whether or not one succeeds in one's new role of chancellor of the exchequer of time. It is as well not to chatter too much about what one is doing, and not to betray a too-pained sadness at the spectacle of a whole world deliberately wasting so many hours out of every day, and therefore never really living. It will be found, ultimately, that in taking care of one's self one has quite all one can do.

Another danger is the danger of being tied to a programme like a slave to a chariot. One's programme must not be allowed to run away with one. It must be respected, but it must not be worshipped as a fetish. A programme of daily employ is not a religion.

This seems obvious. Yet I know men whose lives are a burden to themselves and a distressing burden to their relatives and friends simply because they have failed to appreciate the obvious. "Oh, no," I have heard the martyred wife exclaim, "Arthur always takes the dog out for exercise at eight o'clock and he always begins to read at a quarter to nine. So it's quite out of the question that we should. . ." etc., etc. And the note of absolute finality in that plaintive voice reveals the unsuspected and ridiculous tragedy of a career.

On the other hand, a programme is a programme. And unless it is treated with deference it ceases to be anything but a poor joke. To treat one's programme with exactly the right amount of deference, to live with not too much and not too little elasticity, is scarcely the simple affair it may appear to the inexperienced.

And still another danger is the danger of developing a policy of rush, of being gradually more and more obsessed by what one has to do next. In this way one may come to exist as in a prison, and ones life may cease to be one's own. One may take the dog out for a walk at eight o'clock, and meditate the whole time on the fact that one must begin to read at a quarter to nine, and that one must not be late.

And the occasional deliberate breaking of one's programme will not help to mend matters. The evil springs not from persisting without elasticity in what one has attempted, but from originally attempting too much, from filling one's programme till it runs over. The only cure is to reconstitute the programme, and to attempt less.

But the appetite for knowledge grows by what it feeds on, and there are men who come to like a constant breathless hurry of endeavour. Of them it may be said that a constant breathless hurry is better than an eternal doze.

In any case, if the programme exhibits a tendency to be

oppressive, and yet one wishes not to modify it, an excellent palliative is to pass with exaggerated deliberation from one portion of it to another; for example, to spend five minutes in perfect mental quiescence between chaining up the St. Bernard and opening the book; in other words, to waste five minutes with the entire consciousness of wasting them.

The last, and chiefest danger which I would indicate, is one to which I have already referred--the risk of a failure at the commencement of the enterprise.

I must insist on it.

A failure at the commencement may easily kill outright the newborn impulse towards a complete vitality, and therefore every precaution should be observed to avoid it. The impulse must not be over-taxed. Let the pace of the first lap be even absurdly slow, but let it be as regular as possible.

And, having once decided to achieve a certain task, achieve it at all costs of tedium and distaste. The gain in self-confidence of having accomplished a tiresome labour is immense.

Finally, in choosing the first occupations of those evening hours, be guided by nothing whatever but your taste and natural inclination.

It is a fine thing to be a walking encyclopaedia of philosophy, but if you happen to have no liking for philosophy, and to have a like for the natural history of street-cries, much better leave philosophy alone, and take to street-cries.

後記
人生存摺

· 時間可以創造比金錢更重要的財富。

· 時間更是萬物的根源。

· 時間恰如其分地提供給我們需要，正如生命每天都會
出現奇蹟一般。

· 在一天裡，沒有人可以擁有比任何人更多的時間，也
不會更少。

· 積極有效地使用時間，是迫不及待要學習的事情。

· 時間對所有人都是規律而公平的。

· 一天二十四小時一旦恍恍惚惚地過了，便永遠失去它
的「作用」，亦即流失我們的生命價值。

· 所謂的「生活」，是指我們用什麼態度及方式在經歷
生命的過程。

・事實上，我們永遠不可能有「多出來的時間」。

・欲望或需求也是生命中的一部分。

・我們必須首先確實地瞭解欲望的動機及可能性。

・提升自我的知識水準並滿足求知欲，除了書籍以外，其實還有許多的方式。

・時間的運用根本沒有所謂簡單的方法。

・事實的真相是，不但永遠不可能有「多一點的時間」，我們所擁有的時間就是當下所有的一切。

・不要天真地以為只要認真地寫下一張詳細的計畫表，就是完成人生的規劃。

・人生就是無可預料，順境與逆境往往就在一瞬之間。

・那就「開始」啊！世界上找不到任何方法，可以教我們如何開始過日子的。

・只要我們想有所作為，人生便能隨即重新開始。

・任何事都不可能一開始就順暢，即使是小小的收穫都要感恩地累積成果。

・一兩次挫折在所難免，要相信自己因失敗而成長，並強化自尊心不會輕易被擊倒。

一天始於下班後：
妥善運用你的待機時間

・很多人將工作的早上十點到晚間六點，視為有意義的
「一天」，而早上十點以前的時間及晚間六點以後的時
間，皆為這「一天」的開始及結束。

・只要能妥善地利用這十六個小時，未來所獲得的一切，
都會比每天幻想著從天而降的遺產還要來得多更多。

・只要願意改變態度，我們的人生就會隨之有所變化。

・除了適當而充足的睡眠外，我們的大腦可是充滿幹勁
地需要接受事物不斷的刺激變化，別以為它像四肢一樣
會常常感到肌肉「疲勞」。

・別小看零零碎碎的時間。只要能善加利用，日子一久
後，對我們生命所產生的影響是不容小覷的。

・一日之計在於晨。將鑽石般寶貴的清晨用來看報，尤
其是八卦新聞，還真是白白地浪費掉大好時光。

・在進行時間管理之前，要先調整好自己的心態。首先
應保持平常心，不要操之過急；其次不要輕忽任何渺小
的成就，因為這都可能帶來極大的成功。

・就在你閱報的同時，有許多人已經利用時間，將更多
有意義的時光存進生命帳戶。

．你必須真誠地面對自己。事實上，在你六點下班後的這段時間，你的狀況並沒有如想像中的疲憊不堪。

．你能至少每隔一個晚上用一個半小時來動動腦，活化你被工作快要淹沒的創意力、想想你的人生方向及生活周遭的一切……等。亦即，撥出一段重要的時間來思考生命中重要的事。

．最浪費的時間，是花在想睡又拖拖拉拉不睡的那段零碎時間，重要的是生命在無意義中又消逝了一個小時。

．如果要讓生活有意義，一次學習也不能停止。

．唯有那些長久以來為工作鞠躬盡瘁的人，他們才能瞭解休假的珍貴用意，也才能妥善地利用假日。

．不要以為假日或下班後的時間是必然的所得，你會在漫不經心中就揮霍掉。要將它視為從天而降的禮物，為自己的人生做有意義的計畫及使用。

．千萬別小看每週七小時的成果！重要的是，你務必要來親身體驗一下，這七個小時所能創造的生命奇蹟。

．要改變習慣可不是件輕鬆的事，更別說是在改變的歷程中，經常會出現許多矛盾和衝擊。

．要想讓有益自己的事持之以恆地進行，自信是非常重要的。

．許自己三個月的時間——若是你能確切落實，每週撥出七個小時來自我投資，你就會感受到自己的改變，並發現生活周遭神奇的變化。

．每天的第一項工作，就是要先練習提高集中專注力，意識到自己要開始真正的生活。

．我每天都會將這段時間儲蓄起來——從出家門口的那一刻到抵達公司的那一刻，那是一段培養意志的練習之旅。

．你無法集中注意力，歸根究柢只有一個原因，你太懶散了！

．除了持之以恆外，要完成注意力集中的練習並沒有其他的技巧。

．如果你所想的事是正念而有意義的，那你所得到的好處就更多了。

．當你集中注意力的同時，另一個意外的收穫，那個老是讓你遭受突發性災難的罪魁禍首——恍神，從此也將跟你的人生告別。

‧就如正式學鋼琴前的音階基本課程，每一天你都務必要花三十分鐘以上，學習如何使注意力高度集中。

‧我們必須先學會掌控人體中最複雜、最難以駕馭的器官──大腦，才能進一步供給它有意義的養分（思想）。

‧攸關我們此生的問題，我們都未曾深刻地思考。我們的幸福、我們的未來、生活中的擁有及失去、為何做了這樣的事、為何有那樣的想法⋯⋯等。

‧生活，並非只是活著就好。你要在生活中感受生命存在的價值，也要在生命中體驗有意義的生活。

‧幸福，並非是物質或精神的一時快樂。它是以智慧面對人生，願意相信生命中的一切並全心地為它付出，它使我們的生活獲得永恆的喜樂。

‧建立正確的信念的第一步，最重要的是我們的想法和行為是否一致。

‧實際上，我們在大多數時候的行為卻是受感覺的影響，而非理性的判斷。

‧越不曾經過反省的理智，其功能就會越來越降低。

‧書是無價之寶，但是這並非要用閱讀取代所有的自我訓練，我們仍然要每天誠實反省自己的一切言行。

‧「反省自我」是一件不能忽略的生命功課，你每一天都一定要完成此訓練。

‧不喜歡閱讀的人，其實是沒有找到自己有興趣的書來閱讀，而非他與書天生「無緣」。

‧任何人都有與生俱來品味音樂的能力。

‧品味藝術是件很好的事，但並非是人生中最為重要的事。

‧想要透徹地瞭解生命，就要學會深入觀察事物的因果關係。

‧每件事發生都有其道理。

‧這正如當我們遭逢生命中的逆境時，我們總先本能地怨天尤人；若平心靜氣地回想前因後果，事情的發生果然不是沒有道理。

‧任何事物都有其一體兩面，有惡必有善，重要的是我們是否有正確的信念，而非一味歸咎於劣質的人性。

‧想要避免對人性不必要的恐慌症，化解對生命的無力感及困惑，我們就要加強自己剖析事物的因果由來的能力，及研究改善逆境的解決之道。

‧一個能深悟生命無常及宇宙變化道理的人，與一個只懂得生存的人，他們最大的差別在於瞭解思考的重要性。

‧要學習分析事物的因果關係，其實生活中有許多地方就可以印證。

‧只要知道推斷一件日常事物的方法，同理可證，人生中再複雜的事一樣可解。重點是你願意回歸初心，放下偏見及固執，完整地省視事物。

‧如果你能每天多花個二十分鐘，研讀那些商業鉅子的成功之道，或許你就可以早一天成為財經界的投資天王或天后。

‧只要你願意每天不間斷地花上固定的時數，不論是了自己的工作、生活、家庭，甚至是樂趣，深入而廣大地研究它，不僅可以使你在職場上如魚得水，同時也能有效面對生活。

‧我們不需要浪費時間去閱讀劣質的作品。

．在訓練思考力時，我們必須經常刺激腦子，時時刻刻
保持著對立及質疑的心態，及尋求真理的毅力。

．詩歌，雖然是最讓人腸枯思竭的，但是，它也讓人獲
得最多的智慧及喜悅。

．既然花了二十分鐘的時間看書，那最少也要用四十五
分鐘反覆思量。若是無法從書本中想出些道理，那就無
須要浪費時間閱讀了。

．閱讀時讓自己沉浸於書的世界，就在不經意間，你會
發現自己的人生又到一個美麗新世界。

．安排閱讀時，請先確定研讀的目標及範圍；其次，請
記得邊讀邊思考。

．我們真正能做的、該做的，是盡自己的本分，管好自
己的人生。

．要想真正地落實計畫，就要規劃出自己能力範圍內的
計畫。

．我們經常會訂定太多的計畫，甚至已經超出自己的能
力所及，此時就要懂得節制並減少不必要或緩不濟急的
計畫。

‧失敗會吞噬掉計畫初始的信心與行動，就如同殺害未來會成為大樹的新芽一般。

‧當計畫超出你能力以外，但你又不願改變它時，有一個緩衝的方法是：折衷一部分的時間改做其他的活動或休息。

‧當你計畫要投資人生時，請選擇從自己的興趣開始，其次，一定要與自己的性格相合。

‧進行時間管理時，要首先學會安排自己的時間，其次不要讓自己受限於計畫之中，並且不在匆忙間實行計畫。最後，切莫在計畫實行之初就遭遇太早及太多的失敗，以免喪失信心。

國家圖書館出版品預行編目 (CIP) 資料

一天始於下班後：妥善運用你的待機時間 / 阿諾．班
奈特 (Arnold Bennett) 著 . -- 初版 . -- 新北市：晶冠，
2020.03
面；　公分 . -- (時光薈萃系列；6)
譯自：How to live on 24 hours a day
ISBN 978-986-97438-9-1(平裝)

1. 時間管理 2. 工作效率

494.01　　　　　　　　　　　　　109000046

時光薈萃 06

一天始於下班後：妥善運用你的待機時間

作　　　者	阿諾・班奈特（Arnold Bennett）
繪　　　圖	阿步
行 政 總 編	方柏霖
責 任 編 輯	王逸琦
封 面 設 計	李純菁
出 版 企 劃	晶冠出版有限公司
總 代 理	旭昇圖書有限公司
電　　　話	02-2245-1480（代表號）
傳　　　真	02-2245-1479
郵 政 劃 撥	12935041 旭昇圖書有限公司
地　　　址	235 新北市中和區中山路二段 352 號 2 樓
E-MAIL	s1686688@ms31.hinet.net
旭昇悅讀網	http://ubooks.tw
印　　　製	福霖印刷有限公司
定　　　價	新台幣 250 元
出 版 日 期	2020 年 03 月 初版一刷
ISBN-13	978-986-97438-9-1